COLLINS GEM

CATS

COLLINS GEM

Chinese
ASTROLOGY

COLLINS GEM

Classic
BOOKS

COLLINS GEM

Classic
FILMS

COLLINS GEM

HORSES
& PONIES

COLLINS GEM

INSECTS

COLLINS GEM

**KINGS &
QUEENS**

COLLINS GEM

MUSHROOMS
& TOADSTOOLS

COLLINS GEM

SNAKES

COLLINS GEM

SPIDERS

NS GEM

ROT

COLLINS GEM

WINE
Guide

COLLINS GEM

WORLD
atlas

LINS GEM

ODIAC
Types

COLLINS Jane's

COMBAT
AIRCRAFT

Bob Munro
Christopher Chant

HarperCollins*Publishers*

Picture acknowledgements:
Aviation Photographs International: 40, 42, 178, 210, 323;
Dassault Aviation: 82, 84, 94, 98; Eurocopter: 102-112;
Eurofighter: 114; Jan Kouba: 32; Peter R Foster: 52, 80, 86, 88,
90, 162, 164, 166, 168, 180, 184, 200, 208, 220, 240; Paul
Jackson: 50, 96; Kamov: 132; Lockheed: 140, 142; Bob Munro:
all other photographs; McDonnell Douglas: 148, 150;
Northrop-Grumman: 188; Tim Ripley: 74, 92, 128, 158, 218;
Rockwell International: 194, 196; Saab: 198-204

Thanks to the late Mr William Crampton of the Flag Institute
for the flags

HarperCollins Publishers
PO Box, Glasgow G4 0NB

First published 1995
This edition published 1999
© HarperCollins Publishers 1995
Flags © The Flag Institute

Reprint 10 9 8 7 6 5 4 3 2 1 0

ISBN 0 00 472297-5

All rights reserved. Collins Gem® is a registered
trademark of HarperCollins Publishers Limited

Printed in Italy by Amadeus S.p.A.

Contents

Introduction: Airpower in the 1990s

Until the late 1980s the conduct and development of military aviation was dominated by the Cold War. The advanced industrial nations were divided into two primary camps led by the USA and USSR and separated by the 'Iron Curtain'. Although this long lasting confrontation had been affected by the emergence of China as the core of a third power bloc, and by the not altogether satisfactory extension of the communist/capitalist divide into the so-called 'third world', it held basically true for the most economically advanced countries of the northern hemisphere. Most major air forces adhered either to the communist bloc centred on the Warsaw Treaty Organisation or to the capitalist bloc hinged on the North Atlantic Treaty Organisation.

Cold War legacy

The extreme polarity of this divide and, the continued politico-military antagonism of the two armed camps, despite the onset of détente, made for political and military certainties that were only marginally affected by financial considerations. The major threat in this period was perceived by the West as communist political and territorial ambitions, especially in western Europe, and by the East as capitalist attempts to hem in and strangle communism. Here the cockpit of danger was seen as western Europe in general, and the

A Canadian Air Force CF-18 fires CRV-7 rockets

divided country of Germany, in particular, as this must inevitably serve as the springboard for military adventurism by either side. Military wisdom opined that conventional military operations in Germany, with supporting operations on this country's northern and southern flanks, would almost inevitably be the starting point of World War III, and that impending military defeat on this crucial land/air battlefield would raise the spectre of escalation from conventional to nuclear warfare.

It was these factors that led to certitude in the politico-military machines of each bloc, for the belief

A Soviet MiG-29 seen from a US aircraft in the late 1980s

on each side of the divide was that continued security was wholly dependent on deterrence, in the form of a political willingness and military capability to undertake decisive operations, should the other side launch an attack. Thus there emerged a politico-military culture that emphasized the need to maintain strong forces equipped with the very latest weapons to deter the military ambitions of the other side. This translated into an enormous investment, in both financial and manpower terms, in major defence industries that could undertake vast amounts of advanced research and then turn the results of this

Anglo-French SEPECAT Jaguar strike aircraft of the RAF

research into ever more sophisticated hardware. Thus the politico-military complex, as its opponents liked to call it, could guarantee a good chance of success in any large-scale military operations and, at the same time, provide the domestic advantages of massive employment and the undertaking of the primary research that produced very useful results right across the spectrum of advanced-technology industry, civil as well as military.

In this culture of military excellence with little regard to cost, the development and procurement of advanced military aircraft was seen as essential, for it

9

was aircraft that were seen as the overall arbiters of any World War III: the tank might be decisive on the European land battlefield, but was vulnerable to advanced fixed- and rotary-wing aircraft carrying advanced sensors, computerised nav/attack systems and sophisticated weapons; and the combination of the aircraft carrier and the submarine might be vital to success in the naval arena, but both types were vulnerable to greater or lesser extents to the attentions of fixed- and rotary-wing aircraft. Thus, while each element of warfare might have its own particular weapon, the aeroplane was common to all and therefore worthy of considerable expenditure.

Influence of the 'Star Wars' initiative

This all began to change in the 1980s as President Ronald Reagan of the USA implemented the Strategic Defense Initiative (SDI) that came to be called the Star Wars programme by the popular media. The ultimate threat of modern warfare was inevitably the resort to tactical nuclear weapons and the escalation to strategic nuclear weapons of the type that could wholly devastate the USA. US strategic planners had been compelled to consider the threat to their country posed by air-launched weapons (both bombs and stand-off missiles) since the early 1950s, land-based missiles since the early 1960s and submarine-

The exceptionally agile General Dynamics F-16

launched missiles also since the early 1960s. The threat of a multi-layer attack using all these weapon types had long tasked military planners, resulting in the creation of an air-defence network of manned interceptor fighters and long-range surface-to-air missiles, counter-force missiles, and a combination of nuclear-powered attack submarines, specialised weapons fired from surface ships, and dedicated anti-submarine aircraft of the fixed- and rotary-wing types. What was universally admitted, however, was the impossibility of destroying any but a percentage of these weapons in the event of an all-out nuclear war,

11

Tu-95 'Bears' supported the Soviet navy in the Cold War

and therefore the likelihood of enormous civilian casualties and the economic dislocation, if not outright destruction, of the American economy, even if the USA won the nuclear war.

This was not acceptable to Reagan and his political advisers, and the SDI was conceived as a way of protecting the USA from all but the smallest damage from Soviet nuclear weapons, of which the vast majority would be eliminated long before they could threaten the USA. The way in which this monumental task was to be achieved was the electronic or physical destruction of these weapons while they were still far distant from the USA in the air or in space, and the methods were to reflect the USA's political determination, great wealth and enormous skills in the development and exploitation of the very latest technologies for the creation over the USA of a defensive shield. This would operate autonomously and semi-autonomously on the basis of intelligent computers for the exceptionally rapid detection and prioritization of all threats, and then their destruction, using a variety of extraordinarily advanced, even 'science fiction' weapons.

Technology race

The whole programme pushed the bounds of practicality in terms of financial resources, and went somewhat farther in terms of technology, but did force the USSR into competing in the development of a comparable system. The SDI programme was a severe burden on US financial and research capabilities, but the equivalent Soviet effort was too great for the USSR to bear. With its political and industrial resources already straitened, the USSR could not compete effectively with the USA in this element of the two superpowers rivalry, and its attempt to do so was a key factor in the collapse of the USSR as a political entity in the late 1980s.

The collapse of the USSR had major short-term

implications, and continues to have longer-term effects that are still not fully appreciated. Among the shorter-term results were the dissolution of the USSR into separate countries, now only partially bound into the deeply troubled Commonwealth of Independent States, the end of the Warsaw Pact and the emergence of the USSR's erstwhile satellites in eastern Europe as fully independent countries, and the reunification of the two Germanies into a single federal republic. As far as superpower politics was concerned, the major reality of the USSR's collapse was the end of the so-called Cold War and thus of the genuine threat of global nuclear war — even though this has been replaced by the greater possibility of local nuclear conflicts as the USSR's vast arsenal of tactical as well as strategic nuclear weapons fell into the hands of some of the smaller republics of the erstwhile USSR and, it is rumoured, criminal elements that may be prepared to sell the weapons (or alternatively weapon-grade uranium or plutonium) to countries of acute political paranoia.

The end of the Cold War removed the need for the USA and her allies to maintain large military establishments and supply them with the latest weaponry from a large research and production base. Thus the disappearance of the Soviet military threat has been matched on the other side of the political

The US Navy's F/A-18E/F Hornet has 11 stores stations

US Marine Corps McDonnell Douglas AV-8B Harrier IIs

A Sukhoi Su-17 bomber emerges from its shelter

divide by major reductions in the strength of the armed forces and a massive diminution of spending on military research, development and production.

This origins of this trend were already evident in the 1980s, for increased emphasis on schemes such as the SDI effort meant less money for conventional forces and therefore an increased tendency toward quality rather than quantity. This may in itself be seen as a development of the tendency already apparent in the 1970s, when even the USA was forced to concede that it could no longer afford to design, develop and procure a specific weapon or weapon system for each

The AC-130 Spectre carries a 105 mm gun/howitzer

and every role, and that such a process was not even advisable from the logistical and training points of view. So, while the USA emerged from the Vietnam War in the early 1970s with warplanes that had generally been designed for a single role but then adapted to undertake at least one more major task, the aircraft that began to enter service after that conflict were genuine multi-role types: the McDonnell Douglas F-4 Phantom II had been designed as a fleet defence fighter but then adapted in the Vietnam War as a true multi-role type able to undertake land-based as well as carrierborne roles, but

17

the later McDonnell Douglas F/A-18 Hornet was developed as a type that could undertake either the fighter or attack roles merely by a change of computer software and some small items of external equipment. The combination of the fighter and attack roles in a single airframe meant that US Navy aircraft carriers did not need to embark two different types of role-specialised aircraft with all their separate spares, while the service's procurement and training establishments had a simpler task that also benefited from economies of scale.

Trend towards multi-role aircraft

Throughout the 1980s this tendency toward multi-role capability was emphasized alongside stealthiness, or the quality of low observability that helps to make a warplane as nearly invisible as possible to electromagnetic and infra-red sensors. Multi-role capability was seen to rest primarily with the creation of a basic airframe optimised, not for outright high performance, but for moderately good performance under any and all flight regimes, and fitted with advanced electronics based on a digital computer and databus combination that allowed the integration of ever more sophisticated sensors and weapons without major revision of the airframe. The key elements of the inbuilt electronics were a pulse-Doppler main

The BAe Hawk 100 has sold well in the Middle East

radar providing air-to-surface as well as air-to-air capabilities, a sophisticated nav/attack system based in an inertial navigation system and possessing interfaces for a number of target acquisition and weapon delivery systems, a pilot's head-up display, on which could be projected all vital flight and mission data, and an increasingly sophisticated array of defensive electronics based on various defensive sensors (radar warner, laser warner, missile approach warner etc.) used to trigger ECM pods and/or chaff/flare dispensers for the decoy of radar- and IR-guided missiles respectively.

19

Mirage fighters have been exported all over the World

The need for stealth was revealed most dramatically in the 1973 Yom Kippur War between Israel and her two most dangerous Arab neighbours, Egypt and Syria. The Israelis used mostly modern American warplanes, but found that these could be acquired and tracked by the sensors of the latest surface-to-air weapons systems supplied to the Arab countries by the USSR. Even when the latest American ECM equipment had been hastily added, the Israelis found that these systems degraded but did not entirely remove the ability of the Soviet systems to acquire, track and engage their aircraft, whose shape had been optimised for maximum flight performance rather than relative electromagnetic and thermal invisibility.

Stealth technology

This prompted a major American effort, later copied by other warplane-producing nations, to reduce the observability of their warplanes by designing stealthiness into them. The two approaches adopted for aircraft were faceting and contouring, used individually or collectively. Faceting, as epitomised by the Lockheed F-117 Night Hawk attack warplane, involves the design of the airframe as an angular assembly that reflects electromagnetic radiation in all directions except that back to the transmitter. Contouring, as epitomised by the Northrop B-2 Spirit strategic bomber, requires the design of an airframe that on the outside is highly curvaceous to minimise electromagnetic reflectivity and on the inside is constructed with internal angles and radar-absorbent materials to trap rather than reflect electromagnetic energy. Elements of both concepts are in fact found in each of these aircraft types, which also have propulsion systems that blend cold free-stream air with the hot exhaust gases of their non-afterburning turbofan engines to reduce their thermal and acoustic signatures. Another aspect of emission-reduction is the development of the so-called 'supercruise' capability for supersonic flight without afterburning, and this is a key element in the design of the forthcoming Lockheed/Boeing F-22 Rapier: non-afterburning flight trims the warplane's thermal signature by a considerable margin without sacrificing performance, and also has important benefits in terms of reduced fuel consumption.

Passive detection systems

Other aspects of low observability are a switch away from active systems, such as search and Doppler navigation radars, the beams of which can be detected by passive listening systems, to passive methods such as optronic target-acquisition/weapon-aiming, and inertial and/or terrain-referenced navigation systems. It was also important to reduce the external clutter under the warplane, for this boosted the electromagnetic reflectivity of the warplane to a marked degree. External carriage of disposable weapons, ECM pods, drop tanks and the like, had become increasingly the norm since the later stages of World War II, for external carriage allowed the creation of a smaller and cheaper airframe that was made versatile by its ability to carry an increasingly large and diverse load of external stores under its wings and fuselage. Yet the combination of the airframe, several projecting pylons, multiple ejector

A BAe Harrier FRS.1 banks to reveal its 4 Sidewinders

An A-6 Intruder in US Navy low visibility scheme

racks, and stores of different sizes and shapes produced great electromagnetic reflectivity. In stealth aircraft, therefore, the use of external pylons is avoided in favour of internal weapons accommodation (as on the F-117 and B-2), or external weapons carriage of the conformal type with the carefully shaped weapons carried flush against the airframe for minimum reflectivity. This last tendency is most evident in multi-role warplanes currently under development, most notably machines such as the Eurofighter EF-2000 and Dassault Rafale.

Fly-by-wire controls

These two aircraft and other warplanes such as the F-22 Rapier, Saab JAS 39 Gripen and Sukhoi Su-27 Flanker are based on a design of relaxed static stability, which means that they have to be flown actively the whole time. This is beyond the capabilities of a pilot, and requires the use of a fly-by-wire control system. In this system, the pilot's control inputs are assessed relative to the warplane's attitude and ambient conditions by a computer system, which then generates the control-surface commands which make the warplane respond exactly as the pilot wants without exceeding its structural or aerodynamic limits. The availability of such technology, starting in the 1970s with aircraft such as the General Dynamics

Changing times: a MiG-31 visits a European air show

F-16 Fighting Falcon, McDonnell Douglas F-15
Eagle and Panavia Tornado IDS, but now improved
radically in digital rather than the original analogue
forms, had made current and future warplanes
considerably more agile than their predecessors, and
this tendency is being further enhanced by
aerodynamic/control developments such as control-
configured vehicle technology and the forthcoming
introduction of thrust-vectoring nozzles on aircraft
such as the Sukhoi Su-35 Flanker.

As well as enhancing agility and controllability
through the use of a fly-by-wire (or in more modern

25

Northrop F-5s are flown by many US allies

types a fly-by-light fibre optic) system, the greater use of computers has removed from the pilot a large measure of the housekeeping duties that he had previously to perform just to keep his warplane in the air. Most of these tasks are now undertaken by a computer that informs the pilot only if any parameter strays from nominal. This allows the pilot to concentrate on the tactical aspect of his mission, which are also enhanced by the computer through the head-up display and, increasingly, the replacement of conventional instruments and screens by multi-function head-down displays with monochrome or coloured symbology.

The task now facing manufacturers and air forces is how best to exploit these and other technologies in warplanes suitable for service in a world less threatened by high-intensity global war, but which may see greater demands for medium-intensity capabilities in peacekeeping efforts. These latter are particularly difficult on modern warplanes, for they may require operations from primitive airfields without advanced support facilities in climatic conditions that may vary from extreme heat to bitter cold, and from total aridity to cloying humidity.

Value for money

Thus the emphasis is now placed ever more strongly than before on value for money in terms of versatility, longevity and economy. Versatility is enhanced by the incorporation of modular design concepts, which allow the easy replacement of any damaged component and also the simple upgrade of electronics through the straightforward replacement of an obsolescent system by a more sophisticated system that can be integrated easily via the databus system. Longevity is also improved by this modular approach to design, for it facilitates servicing and repair when required in airframes that are now better designed for long and updatable lives though the use of computer-

A prototype Longbow Apache fires a Hellfire missile

aided design techniques and the adoption of advanced materials such as aluminium/lithium alloys and composite materials in the airframe, and ceramics and single-crystal metals in the engine. Economy is provided by both of the above factors and by a combination of design and manufacturing factors optimised for low purchase cost, low life-cycle cost and long fatigue life.

There is no doubt, therefore, that the modern warplane is well suited to the changed and indeed still changing requirements already thrust upon it and

Their shape makes B-2 bombers hard to detect on radar

likely to be changed in the years to come to meet altering world situations. The modern warplane is capable yet versatile, and provides excellent value for money in terms of cost and long-term capabilities. The most important need now facing manufacturers and operating air forces is to maximise this versatility for effective use under all intermediate- and low-intensity operational conditions without any significant increase in cost and any major detriment to capacity for effective use in high-intensity warfare, should that threat ever re-emerge.

29

Aermacchi MB.339

The MB.339 doubles as trainer and strike aircraft

Embodying typical Italian styling and elegance, the MB.339 trainer/light attack aircraft flew for the first time on 12 August 1978. Designed and built to replace Aermacchi's previous two-seat jet trainer, the MB.326, the new aircraft retained the licence-built Rolls-Royce Viper Mk 632-43 turbojet powerplant; but in other respects it offered significant improvements.

The first production MB.339A took to the air on 20 July 1978, and initial deliveries against an order of 100 examples for the Italian Air Force began in August 1979. Specialized variants in Italian service include the MB.339PAN (mount of the Frecce Tricolori aerobatic team) and the MB.339RM for radio calibration tasks. Up to 2,040 kg (4,500lb) of external stores can be carried by the MB.339A on six underwing stations, an attractive option for some of the smaller air forces that have procured the type.

More recent variants include the MB.339AM for anti-ship operations, its 'bite' coming in the form of two OTO Melara Marte Mk 2A AShMs; the latest variant, the MB.339C, is optimized for use as a lead-in fighter trainer with ground attack capabilities. Development of the MB.339C began in the early 1980s, its first flight taking place on 17 December 1985. Slightly larger than the MB.339A and has a maximum take-off weight of 6,350 kg (14,000lb).

Specification (MB.339C)

Powerplant: one 19.75 kN (4,400 lb st) Piaggio-built Rolls-Royce Viper Mk 680-43 turbojet
Dimensions: length 11.2 m (36 ft 10½ in); height 3.99 m (13 ft 1in); wing span (over tip tanks) 11.22 m (36 ft 9¾ in)
Weights: normal take-off 4,635 kg (10218 lb); MTOW 6,350 kg (13,999 lb)
Performance: max level speed at sea level ('clean') 902 km/h (560 mph); service ceiling 14,630 m (48,000 ft); range 1,965 km (1,221 miles)
Armament: up to 1,814 kg (4,000 lb) comprising AGM-65 ASMs, Marte Mk II AShMs, bombs, rockets, AIM-9P/L AAMs, Magic AAMs, 30mm gun pods, Miniguns, drop tanks and a four-camera recce pod

Aero L-39 Albatros

The L-39 jet trainer is still widely used in eastern Europe

Designed as a replacement for Aero's successful L-29 Delfin, the L-39 Albatros first flew on 4 November 1968 and entered service with the Czech Air Force during 1974. Close to 3,000 have been built, the majority having been supplied to Russia and the former Warsaw Pact countries.

Of all-metal construction, the L-39 is a relatively simple design powered by a single Progress AI-25 TL turbofan engine. A strong undercarriage allows the aircraft to be operated from grass or unprepared runways at take-off weights up to 4,600kg (10,141lb).

Although the L-39's primary role is that of jet trainer, like many in its class it has a weapons capability that allows it to be used as a light attack/point air defence aircraft. An underfuselage gun pod contains a 23mm two-barrel cannon, up to 150 rounds of ammunition being located in the fuselage area immediately above the gun pod; the four underwing hardpoints can carry a variety of combinations of bombs, AAMs (outer pylons only), rocket launchers and munitions dispensers. The inner two pylons are stressed for loads up to 500kg (1,102lb), the outer two for half as much. This offensive capability has been exploited in the L-39ZA, a ground attack/recce version of the L-39ZO weapons trainer (itself a development of the basic L-39C trainer with reinforced wings). A bid to attract export sales from air forces in the West has led to a version of the L-39ZA with Western avionics known as the L-39ZA/MP.

Specification (L-39C)

Powerplant: one 16.87 kN (3,792 lb st) Progress AI-25 TL turbofan
Dimensions: length 12.13 m (39 ft 92 in); height 4.77 m (15 ft 7¼ in); wing span (over tip tanks) 9.46 m (31ft ½ in)
Weights: take-off ('clean') 4,635 kg (10,218 lb); MTOW 5,600 kg (12,346 lb)
Performance: max level speed at sea level 610 km/h (379 mph); service ceiling 7,500 m (24,600 ft)
Armament: (L-39ZA/ART): one 23 mm GSh-23 two-barrel gun with 150 rds; 1,000 kg (2,205 lb) of bombs, rocket launchers, AIM-9 AAMs, drop tanks and training dispensers

Aero L-59

One of the 48 L59s flown by the Egyptian air force

Originally known as the L-39MS, the L-59 is an improved version of the L-39 and first flew on 30 September 1986. Undoubtedly the aim is to win orders in the West, new customers being all the more important since the fall of Communism and the end of all but guaranteed orders on behalf of the air forces of the Soviet Union and its Warsaw Pact allies.

Slightly larger overall than the L-39C, and powered by a single Progress DV-2 turbojet, the L-59 features revised flying controls and upgraded avionics; the latter includes a Bendix-King KNS 660 flight management system and a Flight Visions FV-2000 HUD and mission computer, with a video camera and monitor fitted in the front and rear cockpits respectively. As with the L-39, an underfuselage gun pod houses a 23mm two-barrel GSh-23 cannon, while four underwing hardpoints can carry a variety of ordnance.

Deliveries to the Czech Air Force of the L-59 (first flight 1 October 1989), the initial production version, began in 1991, and an export order for 48 aircraft worth over $200 million was later placed by Egypt. Known as L-59Es, the aircraft for the Egyptian Air Force (a long-time user of the L-39) are fitted with US avionics. A second export order has been placed on behalf of the Tunisian Air Force, for 12 aircraft.

Specification (L-59E)

Powerplant: one 21.57 kN (4,850 lb st) Progress DV-2 turbofan

Dimensions: length 12.20 m (40 ft ¼ in); height 4.77m (15ft 7¼ in); wing span (over tip tanks) 9.54 m (31 ft 3½ in)

Weights: empty, equipped 4,030 kg (8,885 lb); MTOW (with external stores) 7,000 kg (15,432 lb)

Performance: max level speed at 5,000 m (16,400 ft) 865 km/h (537mph); service ceiling 11,800 m (38,725 ft)

Armament: one 23 mm twin-barrel GSh-23 cannon with 150 rds; up to 1,000 kg (2,204 lb) of bombs, 57mm and 130mm rocket pods, gun pods and drop tanks

Agusta A.109

The A.109KM is the military version of the popular A.109

Developed in the 1960s and flown for the first time on 4 August 1971, the Italian A.109 has won over 500 orders to date, mostly for civil versions. During the 1970s Agusta decided to exploit the A.109's military potential and it was no surprise when the Italian Army became the first customer for a scout/anti-tank variant designated A.109EOA. The most successful of the military variants so far has been the A.109CM. In 1988 the Belgian army ordered 46 such machines, comprising 26 helicopters configured for anti-tank operations and 18 configured for scout work. Known as the A.109HA, the anti-tank variant sports a roof-mounted Saab/ESCO HeliTOW 2 sight for use in conjunction with up to eight TOW 2A missiles, these being carried on side-mounted stores attachment points. The unarmed A.109HO scouts sport a roof-mounted Saab Helios stabilizing observation sight.

The A.109KM is another armed variant and can be used on scouting, anti-tank or escort duties. A shipborne version of the A.109KM is designated the A.109KN. As with all military configured A.109s, they can carry a modest but effective ordnance load including gun pods, rocket packs and TOW ATGMs. A degree of self-defence is provided courtesy of the ability to carry Stinger AAMs on external stations and a side-firing, gimble-mounted 7.62/12.7mm gun in the cabin.

Specification (A.109KM)

Powerplant: two 575 kW (771shp) Turbomeca Arriel 1K1 turboshafts

Dimensions: length 11.44 m (37 ft 6 in); height 3.50 m (11 ft 5¾ in); main rotor diameter 11.0 m (36 ft 1in)

Weights: MTOW 2,850 kg (6,238 lb); max slung load (907 kg (2,000 lb)

Performance: (at MTOW): never-exceed speed 281 km/h (175 mph); max rate of climb at sea level 618 m (2,020 ft)/min; service ceiling 6,100 m (20,000 ft); max endurance 4h 16min

Armament: up to eight TOW ATGMs or Stinger AAMs, 70 mm or 80 mm rocket launchers, two 7.62 mm or 12.7 mm gun pods, one cabin-mounted 7.62 mm or 12.7 mm gun

Agusta A.129 Mangusta

The Italian army has ordered 60 A.129 Mangustas

In 1978, some six years after the Italian Army issued a specification for a light anti-tank and scout helicopter, Agusta received the go-ahead for its A.129 Mangusta. The first of five development aircraft took to the air on 11 September 1983.

A tandem two-seat, twin-engined design with a four-blade main rotor capable of withstanding hits from 12.7mm bullets, the A.129 has also been built to withstand hard landings at rates of descent up to 10m (32ft 93/4in) per second. Some 45% of the fuselage structure by weight (excluding engines) is made up of composites, and the widely spaced turboshafts are housed in a fireproof engine compartment. Separate fuel systems and self-sealing/foam-filled fuel tanks reduce the risk of fire still further.

The pilot (rear) and co-pilot/gunner (front) sit in cockpits covered by flat-plate low-glint glazing, their seats having sliding side panels of protective armour. A fully integrated digital multiplex system controlled by two computers handles navigation, flight management and weapons control. A nose-mounted FLIR sensor enables the Mangusta to fight at night, information being displayed to the pilot on a monocle forming part of his IHADSS (integrated helmet and display sighting system). His HIRNS (helicopter infra-red night system) night-vision system allows NOE flight by night.

Specification (A 129)

Powerplant: two 615 kW (825shp) continuous rating Rolls-Royce 1004 turboshafts

Dimensions: length 12.275 m (40 ft 3¼ in); height 2.75 m (9 ft ¼in); width over TOW pods 3.60 m (11 ft 9¼ in); main rotor diameter 11.90 m (39 ft ½ in)

Weights: empty, equipped 2,529 kg (5,575 lb); MTOW 4,100 kg (9,039 lb)

Performance (with eight TOW): max level speed at sea level 250 km/h (155 mph); max rate of climb 618 m (2,028 ft)/min; max endurance, no reserves 3h 5min

Armament: up to 1,200 kg (2,645 lb) comprising eight TOW-2A/HOT/Hellfire ATGMs, AIM-9/Mistral/Stinger AAMs, 52 x 70 mm/81 mm rockets and 7.62/12.7/20 mm gun pods

AIDC AT-3 Tzu-Chung/AT-3A Lui-Meng

Taiwan's first jet trainer, the AT-3 entered service in 1990

Development of the AT-3 Tzu-Chung, Taiwan's first indigenous jet-powered military trainer, began in 1975, the first of two XAT-3 prototypes making its maiden flight on 16 September 1980. Powered by a pair of non-afterburning turbofans housed in prominent nacelles either side of the fuselage, the AT-3 is of tandem configuration with the rear seat elevated by 30cm (12in). Unusually for what is a relatively small aircraft, the AT-3 has an integral weapons bay. Located beneath the rear cockpit area, it offers enough space to house a variety of semi-recessed weapons such as machine-gun packs. The four underwing stations, incorporated in the design to provide a degree of weapons training for student pilots, have enabled the AT-3 to be configured for close air support missions. Approximately 20 AT-3s have been configured for this role, which led to the development of a single-seat ground attack/maritime attack variant. Known as the AT-3A Lui-Meng, it was flown in prototype form after the conversion of one of the two XAT-3s.

The exact status of the AT-3A is uncertain, but it is known that a further variant, the AT-3B, comprises AT-3s retrofitted with the AT-3A Lui-Meng's navigation/attack system. The first flight of a production AT-3 took place on 6 February 1984 and deliveries of 60 production-standard AT-3s to the RoCAF (Republic of China Air Force) began the following month, concluding in January 1990.

Specification (AT-3)

Powerplant: two 15.57 kN (3,500lb st) Garrett TFE731-2-2L non-afterburning turbofans

Dimensions: length (including nose pitot) 12.90 m (42 ft 4 in); height 4.36 m (14 ft 3¾ in); wing span 10.46 m (34 ft 3¾ in)

Weights: empty, equipped 3,855 kg (8,500 lb); MTOW 7,938 kg (17,500 lb)

Performance (at MTOW): max level speed at sea level 898 km/h (558 mph); max rate of climb at sea level 3,078 m (10,100 ft)/min; service ceiling 14,625 m (48,000 ft); endurance (max internal fuel) 3h 12min

Armament: 2,271 kg (6,000 lb) comprising bombs, flare dispensers, rocket launchers, rocket pods, AAMs, machine-gun packs and bomb/rocket training dispensers

AIDC Ching-Kuo

Taiwan's Ching Kuo fighter is now in full production

Although Taiwan has historically bought most of its aircraft from America, the US refusal to supply the Northrop F-20 and the General Dynamics F-16s inspired the IDF (Indigenous Defensive Fighter) programme . Ironically, US companies including General Dynamics, Garrett and Westinghouse assisted and the result was the Ching-Kuo (named after the given names of a former Taiwanese president) single-seat air superiority fighter, first flown on 28 May 1989.

Not dissimilar in appearance to the F-16 and F/A-18 Hornet, the all-metal Ching-Kuo is powered by two licence-built TFE1042-70 afterburning turbofans fed by elliptical air intakes set well back along the fuselage. A multi-mode lookdown/shootdown radar (licence-built Westinghouse AN/APG-67(V) incorporating some elements of the same company's AN/APG-66) offers air and sea search capabilities at ranges up to 150km (93 miles).

A 20mm M61A1 multi-barrel cannon is carried internally in the lower port forward fuselage. Six external hardpoints (two underfuselage, two underwing and two wing-tip) can carry a variety of ordnance. Four prototypes (three single-seater and one two-seater) were constructed, although the second single-seater was subsequently lost during a test flight. These were followed by a 10 pre-production and 60 full-production Ching-Kuo's, all of whch had been delivered by early 1994.

Specification (AIDC Ching Kuo)

Powerplant: two 42.08 kN (9,460 lb st) ITEC (Garrett/AIDC) TFE1042-70 (F125) afterburning turbofans

Dimensions (estimated): length (with nose probe) 14.48 m (47 ft 6 in); wing span (over AAM rails) 9.00 m (29 ft 6 in)

Weights (estimated): internal fuel weight 1,950 kg (4,300 lb); MTOW 9,072 kg (20,000 lb)

Performance (estimated): max level speed Mach 1.7; max rate of climb at sea level 15,240 m (50,000 ft)/min; service ceiling 16,760 m (55,000 ft)

Armament: one 20 mm M61A1 Vulcan cannon; Maverick ASMs, Male Bee II AShMs, bombs, cluster bombs, rocket pods and Sky Sword I/II AAMs

AMX International AMX

The AMX light strike aircraft built by Italy and Brazil

In 1977 the Italian Air Force defined a requirement for a multi-role strike/recce aircraft to supersede its ageing G-91R/Ys and F-104G/Ss. Aeritalia and Aermacchi duly combined to develop the AMX to undertake high-subsonic/low-altitude missions by day/night and in poor visibility. The programme expanded in scope in 1980, when Embraer of Brazil joined up, the Brazilian Air Force having stated a requirement for such an aircraft to replace its AT-26 Xavantes.

The first flight of the AMX took place in May 1984 and initial production aircraft entered service in 1989 with both the Italian and Brazilian Air Forces, aircraft for the latter service being designated A-1 (single-seat) and TA-1 (two-seat). Powered by a licence-built Rolls-Royce Spey RB.168 non-afterburning turbofan, the AMX is a compact and conventional design with a shoulder-mounted wing, and is capable of operating from unprepared or partially-damaged runways. External ordnance and auxiliary fuel tanks can be carried on a total of five stations (four underwing, one centreline), while the two wing-tip rails can carry AAMs to provide a degree of self-defence.

Current Italian/Brazilian requirements call for 192 aircraft to be built.

Specification (AMX)

Powerplant: one 49.1 kN (11,030 lb st) Rolls-Royce Spey Mk 807 non-afterburning turbofan

Dimensions: length 13.23 m (43 ft 5 in); height 4.55 m (14 ft 11¼ in); wing span (over AAMs) 9.97 m (32 ft 8½ in)

Weights: operational, empty 6,730 kg (14,837 lb); MTOW 13,000 kg (28,660 lb)

Performance (at MTOW): max level speed Mach 0.86; max rate of climb at sea level 3,124 m (10,250 ft)/min; service ceiling 13,000 m (42,650 ft)

Armament: one M61A1 multi-barrel 20 mm cannon with 350 rds (AMX) or two DEFA 544 30 mm cannon (A-1); up to 3,800 kg (8,377 lb) comprising free-fall/retarded bombs, LGBs, cluster bombs, ASMs, AShMs, PGMs, rocket launchers and AAMs

Atlas Cheetah

The Cheetah benefited from Israeli experience with the Kfir

At first glance, it would be quite easy to mistake the Cheetah for the Israeli Kfir (qv), both designs having canard foreplanes and nose-mounted strakes added to the distinctive delta-winged lines of the highly successful Dassault Mirage III. In fact, Israeli technology and experience with the Kfir played an influential part in helping Atlas to bring the Cheetah project to fruition at a time when South Africa was the subject of an international arms embargo.

The first example was rolled out in July 1986. Designated Cheetah DZ, it was a converted and upgraded Mirage IIID2Z two-seater, one of eight to be similarly modified in due course. Declared operational within the SAAF in October 1987, the

Cheetah DZ force has been expanded with the addition of at least four converted two-seat Mirage IIIDZs. These were followed by 21 single-seater conversions known as Cheetah EZs, based on Mirage IIIEZs (14), IIIRZs (four) and IIIR2Zs (three).

In addition to the canards, the Cheetah DZs sport a longer, downturned nose (again similar to the Kfir TC2 trainer and its derivatives), housing a multi-mode lookdown search and track radar and extra avionics that give these two-seaters a combat capability. The Cheetah EZs have a less bulbous nose, but a fuselage plug has been inserted immediately ahead of the windscreen, the extra space allowing room for an Elta ranging radar and RWR antennae. Cockpit features include an Israeli HUD and nav/attack system and a South African pilot's helmet sighting system.

Specification (Cheetah EZ)

Powerplant: one 70.82 kN (15,873 lb st) SNECMA Atar 9K-50 afterburning turbojet

Dimensions: length (including probe) 15.65 m (51 ft 4¼ in); height 4.55m (14 ft 11¼ in); wing span 8.22 m (26 ft 11 in)

Performance: max level speed ('clean') 2,338 km/h (1,453 mph); service ceiling 17,000 m (55,775 ft)

Armament: two DEFA 30 mm cannon; up to 4,000 kg (8,818 lb) of AS30 ASMs, LGBs, CBUs and AAMs

Atlas CSH-2 Rooivalk

Rooivalks being demonstrated for the British Army

The international arms embargo against the apartheid regime in South Africa failed to prevent South African industry from developing first class military equipment of its own. The SAAF (South African Air Force) issued its first contracts for a helicopter gunship in 1981. First to fly was the XH-1 Alpha cockpit/gun turret test-bed (based on the Alouette III utility helicopter), which took to the air in 1985. This was followed by the XTP-1: a much more ambitious project based on the SA 330L Puma, also in service with the SAAF, which has led to a dedicated Puma gunship. More important was its contribution to South Africa's definitive attack helicopter, the CSH-2 Rooivalk (Red Kestrel). This impressive machine features stepped cockpits covered by low-glint transparencies, high-absorption main landing gear, IR heat suppressors on each of the Topaz turboshaft engines' exhausts, and main and tail rotor systems whose blades are made up of composites. A target detection and tracking system utilizing FLIR, LLTV and a laser rangefinder is mounted in a gyrostabilized nose turret beneath which is a chin-mounted 20mm or 30mm gun.

First flown in prototype (XH-2) form on 11 February 1990, test and evaluation of the Rooivalk is underway using Experimental Development Model and Advanced Demonstration Model prototypes. A third, Engineering Development Model prototype will join the programme in 1996.

Specification (CSH-2)

Powerplant: two 1,491 kW (2,000shp) Topaz (locally upgraded Turbomeca Makila 1A1) turboshafts

Dimensions: length (overall, rotors turning) 18.731m (61ft 5¹/₂ in); height overall 5.187 m (17 ft ¼ in); wing span (over AAMs) 5.198 m (17 ft ¾ in)

Weights: empty 5,910 kg (13,029 lb); MTOW 8,750 kg (19,280 lb)

Performance (at 7,500 kg (16,535 lb) combat weight): never-exceed speed 309 km/h (192 mph); max rate of climb 670 m (2,200 ft)/min; service ceiling 6,100 m (20,000 ft); endurance (at MTOW with external fuel) 7h 22min

Armament: one Kentron GA-1 Rattler/Armscor MG 151 20 mm gun, or one 30 mm gun; ZT-3 Swift/ZT-35 ATGMs, 68 mm rocket launchers and V3C Darter AAMs

Avioane IAR-99 Soim/IAR-109 Swift

Israeli avionics are being offered on the latest IAR-99s

Established in 1972 as IAv Craiova, but renamed Avioane in 1991, this young Romanian company has produced an advanced jet trainer and light ground attack aircraft. Developed during the late 1970s, the IAR-99 Soim (Hawk) was revealed to the West in 1983. Three prototypes (one for structural fatigue testing) were built, the first one making its maiden flight on 21 December 1985. A smart, simple design of all-metal construction, the IAR-99's role as a trainer is enhanced by twin hydraulically actuated airbrakes beneath the rear fuselage, and single-slotted wing flaps which retract when the aircraft's airspeed reaches 300km/h (186mph). Those aircraft configured for ground attack duties have an electrically controlled gyroscopic gunsight and gun camera, as well as the weapons release and gun-firing controls, in the front cockpit. Separate canopies are fitted, whereas IAR-99 trainers have a one-piece canopy.

Twenty IAR-99s have been acquired by the Romanian Air Force and another 30 are in the process of being built. Avioane would seem to be pinning its hopes for future sales success on the IAR-109 Swift, an upgraded IAR-99 announced in 1992. The close-support version (IAR-109TF) has a HUD and laser rangefinder.

Specification (IAR-99)

Powerplant: one 17.79 kN (4,000 lb st) Turbomecanica (Romanian-built) Rolls-Royce Viper Mk 632-41M non-afterburning turbojet

Dimensions: length 11.09 m (36 ft 1¹/₂in); height 3.89 m (12 ft 9½ in); wing span 9.85 m (32 ft 3¾ in)

Weights: empty, equipped 3,200 kg (7,055 lb); MTOW 5,560 kg (12,258 lb)

Performance: max level speed at sea level 865 km/h (537 mph); max rate of climb at sea level 2,100 m (6,890 ft)/min; service ceiling 12,900 m (42,325 ft); max endurance 1h 46min

Armament: removable GSh-23 23 mm ventral gun pod with 200rds; up to 1,000 kg (2,204 lb) of bombs, L16-57 57 mm rocket launchers, 42 mm rocket launchers, two 7.62 mm machine-gun pods, AAMs and fuel tanks

BAe Hunter

After 45 years, the Hunter still flies with a few air forces

First flown in July 1951 as the Hawker P.1067 prototype and entering service with the RAF in July 1954 as the Hunter F.Mk 1 interceptor with an armament of four 30mm cannon, the Hunter was the first swept-wing transonic fighter of British design to enter service. The overall soundness of the design is attested by the fact that the type is still in limited, but useful service with a number of European, South American and Far Eastern countries.

The basic Hunter interceptor went though a number of variants in the interceptor role with a powerplant of either the Rolls-Royce Avon or Armstrong Siddeley Sapphire axial-flow turbojet before reaching its definitive forms as the Avon-powered Hunter FGA.Mk 9 ground-attack fighter and Hunter FR.Mk 10 tactical reconnaissance fighter. The Hunter FGA.Mk 9 was introduced in 1959 with strengthened landing gear, a braking chute, and provision for a larger and more diverse load of underwing stores. The Hunter FR.Mk 10 was similar to the Hunter FGA.Mk 9, except in the forward fuselage, where a fan of three cameras replaced the basic fighter's ranging radar. The most potent surviving model is Switzerland's Hunter Mk 58 with a Saab weapon system and provision for the AGM-65 Maverick air-to-surface and AIM-9 Sidewinder air-to-air missiles.

Several countries still operating the single-seat Hunter also fly small numbers of side-by-side two-seat trainers, derived from the basic Hunter T.Mk 7, as conversion trainers. Some single-seat Hunters retired from first-line service are now employed as weapons trainers.

Specification (Hunter FGA.Mk 9)

Powerplant: one 45.15 kN (10,150 lb st) Rolls-Royce Avon RA.27 Mk 207 turbojet

Dimensions: length 13.98 m (45 ft 10½ in); height 4.02 m (13 ft 2in); wing span 10.25 m (33 ft 8 n)

Weights: take-off ('clean') 8,165 kg (18,000 lb); MTOW 11,158 kg (24,600 lb)

Performance: max level speed at sea level Mach 0.93 or 1,144 km/h (710 mph); service ceiling 15,700 m (51,500 ft)

Armament: four 30 mm Aden Mk 4 cannon with 135 rounds per gun; 2,268 kg (6,000 lb) of bombs, rocket launchers, napalm tanks and drop tanks carried on four underwing hardpoints

British Aerospace Harrier GR.7

The RAF's Harrier GR.7s are designed to attack at night

Unsuccessful attempts by BAe to develop a next-generation Harrier during the late 1970s led instead to collaboration with McDonnell Douglas of the USA. This resulted in the AV-8B Harrier II which first took to the air on 9 November 1978. Initial production of the AV-8B was for the USMC, but the RAF soon followed with an order for 62 aircraft, designated Harrier GR.5s. Deliveries began on 1 July 1987, but it was already clear that they would be superseded by a night attack-capable variant designated the GR.7. Upgrading of all extant GR.5s was completed by 1994, these joining 34 new-build GR.7s in squadron service, the first having been delivered May 1992.

Like the AV-8B Harrier II, the RAF's Harrier GR.7

features a longer fuselage, a 20% bigger wing with outriggers repositioned to mid-span and larger trailing-edge flaps. The redesigned forward fuselage includes a raised cockpit and reconfigured engine air intakes. Optimized for the night attack role, the GR.7 features an NVG-compatible cockpit featuring a digital colour moving map display and wide-angle HUD/HDD. Primary weapons delivery sensors comprise an ARBS with TV and laser target seeker/tracker in the nose, and a FLIR in the prominent fairing atop the nose. Defensive systems include Zeus ECM, and chaff/flare dispensers in the rear of the two AIM-9 AAM launch rails.

Specification (Harrier GR.7)

Powerplant: one 96.75 kN (21,750 lb st) Rolls-Royce Pegasus Mk 105 turbofan

Dimensions: length 14.36 m (37 ft 1½ in); height 3.55 m (11 ft 7¾ in); wing span 9.25 m (30 ft 4 in)

Weights: empty 7,050 kg (15,542 lb); maximum take-off (STO) 14,061 kg (31,000 lb); maximum take-off (VTO) 8,505 kg (18,950 lb)

Performance: maximum level speed ('clean' at 10,975 m (36,000 ft)) 967 km/h (601 mph); 'clean' at sea level) 1,065 km/h (661mph); take-off run (at MTOW) 405 m (1,330 ft)

Armament: two Aden 25 mm revolver cannon in ventral pods with 100rds each; up to 4,173 kg (9,200 lb) of bombs, cluster bombs, 68mm rocket pods, CRV-7 rockets and AIM-9 AAMs

British Aerospace Hawk 100

The Hawk 100 trainer has a useful combat capability

Although the primary role of the Hawk is that of advanced jet trainer for student pilots, a modest weapons-carrying capability has increased its versatility. (The RAF T.1s have three stores points for tactical weapons, and 88 T.1As carry a ventral 30mm gun pod and a pair of underwing AIM-9Ls). But it was not until BAe announced the Hawk 100 that the Hawk's combat potential was seriously exploited.

The first full production prototype Hawk Mk 102D was completed and test flown in February 1992. Key features include a new combat wing with fixed leading-edge droop to enhance manoeuvrability and lift, wing-tip AAM rails and four underwing stores stations, full-width flap vanes and manually operated combat flaps, a longer reprofiled nose housing optional FLIR and/or laser ranging optics, a taller fin, and an upgraded weapons management system. Fuel is carried in one 832l (183gal) fuselage bag tank and two 412.5l (90.5gal) integral wing tanks.

Auxiliary fuel tanks can also be carried on each inner underwing pylon, and reconnaissance tasks can be facilitated by a recce pod mounted on the centreline station in place of the 30mm gun pod or any other weapon.

Not surprisingly, air forces already operating Hawks in the training role have been among those to order the Hawk 100. Customers include Abu Dhabi, Brunei, Indonesia, Malaysia, Oman and Saudi Arabia.

Specification

Powerplant: one 26.0 kN (5,845 lb st) Rolls-Royce/Turbomeca Adour Mk 871 non-afterburning turbofan

Dimensions: length 11.68 m (38 ft 4 in); height 3.99 m (13 ft 1¼ in); wing span (over AAMs) 9.94 m (32 ft 8¾ in)

Weights: empty 4,400 kg (9,700 lb); MTOW 9,100 kg (20,061 lb)

Performance: maximum level speed 'clean' at 10,975 m (36,000 ft) 1,039 km/h (645 mph); maximum rate of climb a sea level 3,597 m (11,800 ft)/min; service ceiling 13,545 m (44,500 ft); endurance approx 2h 6min

Armament: one Aden Mk 4 30 mm ventral cannon with 120rds; up to 3,000 kg (6,661 lb) of ordnance

British Aerospace Hawk 200

The Hawk 200 doubles as trainer and light strike aircraft

By far the most visually distinctive model of the Hawk family is the Hawk 200 single-seat multi-role combat aircraft. First flown on 19 May 1986, this new variant was designed to capitalize on the sales success of the Hawk trainer. Consequently, 80% of its airframe components are compatible with the Hawk Mk 60 and 100.

The major area of redesign involves the forward fuselage and nose. The single-seat cockpit includes a GEC-Marconi MFD, a combined comm/nav interface enabling the pilot to control all such functions from one panel. The more bulbous nose houses a Westinghouse AN/APG-66H multi-mode radar. Other systems include IFF, INS, an optional HUDWAC and RWR. An IFR probe can be fitted next to the windscreen on the starboard side of the fuselage.

The Hawk 100's combat wing has been retained, thus giving the Hawk 200 a highly credible stores-carrying ability - an important feature given the emphasis placed on the aircraft's multi-role capabilities. Surprisingly, no internal armament is fitted, the 30mm cannon having been deleted early in the development programme. Like the Hawk 100, four underwing and one centreline stores pylons are available, plus two wing-tip rails for self-defence AAMs. All pylons are cleared for manoeuvres up to 8g while carrying loads up to 500kg (1,102lb). The Hawk 200 has won several orders from air forces already using the Hawk as an advanced jet trainer.

Specification

Powerplant: one 26.0 kN (5,845 lb st) Rolls-Royce/Turbomeca Adour Mk 871 non-afterburning turbofan

Dimensions: length 11.34 m (37 ft 2½ in); height 4.133 m (13 ft 6¾ in); wing span (over AAMs) 9.94 m (32 ft 7½ in)

Weights: basic empty 4,450 kg (9,810 lb); MTOW 9,100 kg (20,061 lb)

Performance (estimated): max speed at sea level 1,065 km/h (661 mph); max rate of climb at sea level 3,508 m (11,510 ft)/min; service ceiling 13,720 m (45,000 ft)

Armament: up to 3,000 kg (6,614 lb) of bombs, cluster bombs, rockets and AAMs

British Aerospace Nimrod MR.2

Nimrods flew long range recce missions to the Falklands

Based on the de Havilland Comet 4C airliner, the Hawker Siddeley HS.801 maritime reconaissance aircraft combined the former's wings and fuselage with four Rolls-Royce Spey turbofan engines. In addition, an all-new ventral section was added beneath the existing fuselage to create a long internal weapons bay. Two Comets were used as prototypes for the new aircraft, subsequently christened Nimrod MR.1, other features including a fin-tip 'football' containing ESM equipment and a prominent tail-mounted MAD 'stinger'. The first prototype made its maiden flight on 23 May 1967, followed some 11 months later by the first Nimrod MR.1. Forty-six MR.1s were ordered by the RAF, the first of which entered service during 1969.

Of the 46 MR.1s ordered, 35 were upgraded to MR.2 standard, the principal features being the addition of Searchwater radar, an inertial navigation system, separate data processors and a new computer for navigation, acoustics and radar as part of a new central tactical system, and new communications equipment.

The first Nimrod MR.2 was delivered to the RAF in August 1979. Less than three years later the MR.2 fleet was heavily involved in the Falklands War, the addition of an IFR probe resulting in a change of designation to MR.2P.

Today the Nimrod fleet is the subject of much speculation concerning a possible replacement in the early 21st century.

Specification (Nimrod MR.2P)

Powerplant: four 54.0 kN (12,140 lb st) Rolls-Royce RB.168-20 Spey Mk 250 non-afterburning turbofans

Dimensions: length 38.63 m (126 ft 9 in); height 9.08 m (29 ft 8½ n); wing span 35.00 m (114 ft 10 in)

Performance: maximum cruising speed at optimim altitude 880 km/h (547 mph); typical patrol speed at low-level, on two engines 370 km/h (230 mph); service ceiling 12,800 m (42,000 ft); maximum endurance 15 hours

Armament: up to 6,124 kg (13,500 lb) of bombs, depth charges, Harpoon ASMs, Stingray torpedoes and AIM-9 AAMs.

British Aerospace Sea Harrier FRS.1

In 1982 FRS.1s shot down 22 Argentinian aircraft

That the Sea Harrier is a proven combat aircraft is beyond doubt, yet its development was delayed by successive British governments. Better late than never, the prototype Sea Harrier FRS.1 finally took to the air on 20 August 1978.

Based on the RAF's Harrier GR.3, the Sea Harrier FRS.1 introduced a revised forward fuselage incorporating a repositioned cockpit featuring a raised seat, revised canopy and more internal space for an enhanced nav/attack/combat system, and a nose deepened to house a Blue Fox multi-mode radar offering air-to-air and air-to-ground modes. The nose was designed to hinge 180 degrees to port, thus making the most of the limited space below deck when at sea. All magnesium components (susceptible to corrosion by salt water) were replaced,

The first operational FRS.1s joined No.899 Squadron, Fleet Air Arm in April 1990; two years later 29 Sea Harriers flew a total of 2,376 sorties, downing 22 Argentinian aircraft in the process for the loss of six of their own (two in ground accidents) during the Falklands War.

The Fleet Air Arm procured a total of 57 FRS.1s, in-service upgrades including an improved radar, twin AAM launchers and larger drop tanks. All extant FRS.1s are in the process of being reworked to F/A.s standard (qv), the programme having started in October 1990. The sole export customer was the Indian Navy, with a total of 24 FRS.51s and four T.60 trainers.

Specification

Powerplant: one 95.6 kN (21,500 lb st) Rolls-Royce Pegasus Mk 104 vectored thrust non-afterburning turbofan

Dimensions: length 14.50 m (47 ft 7 in); height 3.71m (12 ft 2 in); wing span 7.70 m (25 ft 3 in)

Weights: operating, empty 6,374 kg (14,052 lb); MTOW 11,880 kg (26,200 lb)

Performance: max level speed at low-level 1,185 km/h (736 mph); crusing speed at low-level 650-833 km/h (404-518 mph); STO run at MTOW (no ski-jump) 305 m (1,000 ft)

Armament: two optional 30 mm Aden gun pods; up to 3,630 kg (8,000 lb) of WE177 free-fall/retarded nuclear bombs, Sea Eagle AShMs, AIM-120 AMRAAMs and AIM-9 AAMs

British Aerospace Sea Harrier F/A.2

Sea Harrier F/A.2s are fitted with Blue Vixen radar

In 1985 BAe was awarded the contract for a mid-life update of the Royal Navy's FRS.1 Sea Harriers. Two converted FRS.1s acted as prototypes, the first taking to the air on 19 September 1988. In 1994 the upgraded aircraft was designated the F/A.2.

The most obvious difference between Sea Harrier FRS.1 and F/A.2 is the latter's deeper, reshaped nose radome, necessary to house the GEC-Marconi Blue Vixen radar. Capable of multiple target engagement, Blue Vixen also offers an increase in AAM launch range and better acquisition of surface targets. A redesigned cockpit includes the relocation of important weapon system controls to the up-front control panel or the HOTAS control column, and new displays for the pilot include dual multi-purpose HDDs to complement the existing HUD.

Fitment of the more capable Blue Vixen radar in place of the FRS.1's Blue Fox significantly enhances the F/A.2's combat capabilities, a fact reflected in the new primary armament of up to four AIM-120 AMRAAMs. One such missile can be carried on each outboard launcher rail along with an AIM-9L/M on a side-mounted ejector rail for short-range work. An additional pair of AMRAAMs can be carried in place of the more familiar ventral gun pods. Short take-off air-to-surface attack sorties can be conducted with up to 3,630kg (8,000lb).

Specification

Powerplant: one 95.64 kN (21,500 lb st) Rolls-Royce Pegasus Mk 106 vectored thrust non-afterburning turbofan

Dimensions: length 14.17 m (46 ft 6 in); height: 3.71 m (12 ft 2 in); wing span 7.70 m (25 ft 3 in)

Weights: empty, operating 6,374 kg (14,052 lb); MTOW 11,884 kg (26,200 lb)

Performance: maximum level speed ('clean' at sea level) +1,185 km/h (736 mph); service ceiling 15,545 m (51,000 ft)

Armament: two Aden 30 mm ventral gun pods; up to 3,630 kg (8,000 lb) of WE177 free-fall/retarded nuclear bombs, BL755 and 454 kg (1,000 lb) conventional bombs, AIM-120 AMRAAMs, ALARMs and Sea Eagle AShMs

British Aerospace Strikemaster

Widely exported, the Strikemaster has served since 1967

Adopted as a two-seat trainer by the RAF, the Hunting Jet Provost was soon developed into a light strike/reconnaissance aircraft. The BAC.145 developed from the pressurized Jet Provost T.5 led to the more powerful BAC.167 Strikemaster first flown on 26 October 1967.

The Strikemaster's airframe is strengthened for close-support missions. Thanks to the short-stroke landing gear, operations are possible from unprepared airstrips - a factor that has made the Strikemaster an attractive option for several air forces in Africa and the Middle East. Fuel is carried in the wings in integral and bag tanks, as well as the two fixed wing-tip tanks. In the cockpit the two-man crew sit on side-by-side PB4 ejection seats.

For what is a relatively small aircraft, the

Strikemaster can pack a very respectable punch, carrying up to 1,360kg (3,000lb) of ordnance on eight underwing stores hardpoints. In addition, a pair of forward-firing 7.62mm machine guns (each with 550 rounds of ammunition) are fitted internally, one each buried in the bottom of the lateral engine air intakes.

The Strikemaster enjoyed considerable export success before production ended in 1988; the final six new-build aircraft being supplied to the Ecuadorian Air Force . Total production amounted to 151 aircraft, and although the next generation of combat-capable trainer designs (such as the BAe Hawk and Aermacchi MB.339) have been acquired as replacements by some Strikemaster operators, a respectable number remain in service.

Specification

Powerplant: one 15.47 kN (3,410 lb st) Rolls-Royce Viper Mk 535 non-afterburning turbojet

Dimensions: length 10.27 m (33 ft 8½ in); height 3.34 m (10 ft 11½ in); wing span 11.23 m (36 ft 10i n)

Weights: empty 2,810 kg (6,195 lb); MTOW 5,216 kg (11,500 lb)

Performance: maximum speed ('clean' at 5,485 m (18,000 ft)) 774 km/h (481 mph); initial rate of climb 1,600 m (5,250 ft)/min; service ceiling 12,190 m (40,000 ft)

Armament: two FN 7.62 mm machine-guns with 550rpg; up to 1,361 kg (3,000 lb) of ordnance/drop tanks

Bell AH-1W SuperCobra

The US Marine Corps operates 42 SuperCobra gunships

Although the AH-1 HueyCobra was developed for the US Army, it was not long before US Marine Corps interest in the type led to a firm order for 46 AH-1J SeaCobras. Similar in many respects to the Army's AH-1G, the Sea Cobra was powered by a T400-CP-400 twin-turbine powerplant, and was joined by 38 ex-Army AH-1Gs for training purposes. A further 20 AH-1Js were acquired in the mid-1970s, the final two of which acted as prototypes for the next USMC model, the AH-1T Improved SeaCobra.

Forty-two of the 57 AH-1Ts built were subsequently upgraded to AH-1W SuperCobra standard, the most obvious differences being the two T700-GE-700 turboshaft engines either side of the upper mid-fuselage. Initially designated AH-1T+ and powered by the earlier T700-GE-401 powerplant, the twin-engined model first flew on 16 November 1983. A longer fuselage incorporates extra fuel in self-sealing rubber cells resistant to 12.7mm ammunition, and armour protection is provided for the two-man crew. Each cockpit features dual controls and a HUD compatible with night vision goggles. Planned procurement of the AH-1W by the USMC has been cut from 190 to 154, but export orders have been placed by Turkey and Taiwan. A modified version, the Venom has been offered to the British Army.

Specification (AH-1W SuperCobra)

Powerplant: two 1,285 kW (1,723 shp) General Electric T700-GE-401 turboshafts

Dimensions: length, rotors turning, 17.68 m (58 ft 0 in); height overall 4.44 m (14 ft 7 in); wing span 3.28 m (10 ft 9 in)

Weights: empty 4,634 kg (10,216 lb); MTOW 6,690 kg (14,750 lb)

Performance (MTOW): never-exceed speed 352 km/h (219 mph); max level speed at sea level 282 km/h (175 mph); rate of climb at sea level 244 m (800 ft)/min; service ceiling +4,270 m (14,000 ft)

Armament: one undernose M197 three-barrel 20 mm gun; up to 1,119 kg (2,466 lb) of ATGMs, bombs or AAMs

Bell OH-58D Kiowa/Kiowa Warrior

Used in Vietnam, the OH-58 now doubles as a gunship

In the mid-1960s, Bell developed a military version of the 206A JetRanger. Some 2,200 were acquired by the US Army as OH-58A Kiowas and another 150 or so were procured by overseas customers. A programme to upgrade 585 OH-58As led to the OH-58C, features including a flat glass canopy, IR suppression, better avionics and an uprated engine. Further development of the OH-58 family started in the early 1980s, after Bell's Model 406 won the US Army's competition for a new scout helicopter. The resultant OH-58D introduced a mast-mounted sight housing TV and IR optics and a laser designator/rangefinder, a four-blade main rotor, a new cockpit control and display subsystem and revised avionics.

The OH-58D has in turn lent itself to further development. First came 15 Prime Chance OH-58Ds: hastily converted machines armed for operations against Iranian patrol boats in the Persian Gulf in 1987. Their weapons-carrying abilities were duly incorporated in the OH-58D Kiowa Warrior. Defensive avionics include RWR, an IR jammer and a laser warning receiver.

Current US plans call for 243 unarmed Kiowas to be upgraded to Kiowa Warrior status for US Army use, while 81 Kiowa Warriors will themselves be modified to MPLH configuration, the main features of which are quick-fold rotor blades and a tilting fin to enhance air transportability and rapid deployment.

Specification (OH-58D Kiowa Warrior)

Powerplant: one 485 kW (650 shp) Allison 250-C30X (T703-AD-700) turboshaft

Dimensions: length (overall, rotors turning) 12.85 m (42 ft 2 in); height overall 3.93 m (12 ft 10½ in); width (rotors folded) 1.97 m (6 ft 5½ in)

Weights: empty 1,492 kg (3,298 lb); MTOW 2,495 kg (5,500 lb)

Performance: (MTOW) never-exceed speed 241 km/h (149 mph); max level speed at 1,220 m (4,000 ft) 204 km/h (127 mph); max rate of climb at sea level 469 m (1,540 ft)/min; service ceiling 4,575 m (15,000 ft); endurance 2h 24min

Armament: up to 907 kg (2,000 lb) of ordnance/weapon pods

Boeing B-52G/H Stratofortress

Late model B-52s were converted to carry cruise missiles

Now in the twilight of its operational career, the mighty B-52 has served the USAF faithfully since June 1955. Current plans foresee surviving B-52Hs still in service at the turn of the century, and possibly still around to celebrate the 'BUFF's' 50 years of service.

Originally conceived as a turboprop replacement for Boeing's B-50, the B-52 emerged in 1952 powered by J57 turbojets. Production models ran from the B-52A to the B-52H, each model introducing various improvements and upgrades along the way. Today, only the B-52G (193 built) and B-52H (102 built) remain in service, with the final examples of the former due for imminent retirement. Improvements introduced on the

B-52G included greater fuel capacity, a shorter fin, and a redesigned airframe to increase safety and reduce weight. Originally intended as a platform for the AGM-28 Hound Dog missile, as time progressed the G-model fleet found itself operating in the stand-off penetration role armed with cruise missiles or in a maritime role armed with AGM-88 Harpoons. Low-level all-weather operations by night are possible thanks to a terrain-avoidance radar, FLIR and LLLTV sensors.

The B-52H has a strengthened airframe for low-level operations, and a six-barrel 20mm radar-directed gun replaces the four 12.7mm tail guns of earlier models .

Specification (B-52H)

Powerplant: eight 75.62 kN (17,000 lb st) Pratt & Whitney TF33-P-3 turbofans

Dimensions: length 49.05 m (160 ft 11 in); height 12.40 m (40 ft 8 in); wing span 56.39 m (185 ft 0 in)

Weights: MTOW 229,088 kg (505,000 lb)

Performance: cruising speed at high altitude 819 km/h (509 mph); penetration speed at low altitude 652-676 km/h (405-420 mph); service ceiling 16,765 m (55,000 ft); range +16,093 km (10,000 miles)

Armament: one Vulcan 20 mm cannon in tail turret; up to 22,680 kg (50,000 lb) of AGM-86C ALCMs, B61/83 nuclear weapons, AGM-142 Have Nap PGMs and 51 x 340 kg (750 lb)/454 kg (1,000 lb) conventional bombs

Boeing/Sikorsky RAH-66 Comanche

The Commanche is scheduled to enter production in 2001

The futuristic RAH-66 Comanche won the 1980s ompetition to select and develop a replacement for the thousands of AH-1, OH-6, OH-58 and UH-1s helicopters in US Army service. Designs from Boeing/Sikorsky and Bell/McDonnell Douglas were evaluated, the former being selected on 5 April 1991 and subsequently designated the RAH-66 Comanche. Three YRAH-66 dem/eval prototypes have been funded, with first flight scheduled for August 1995.

Composites make up a large proportion of the Comanche's airframe, as well as the five-blade main rotor and eight-blade fan-in-fin tail rotor. The tandem cockpit arrangement has the pilot in front and weapons operator behind - the reverse of most current tandem cockpit attack helicopters. A FLIR system and laser designator enable the Comanche to fly and fight by night and it will carry a version of the radar developed for the AH-64D Longbow Apache.

Beneath and behind the sensor nose turret is a three-barrel 20mm cannon housing. Further back, behind the main undercarriage units, a pair of side-opening weapons bays can each house three Hellfire ATGMs or six Stinger AAMs. Immediately above these bays, optional stub-wings can carry a further four Hellfires or eight Stingers at each tip. Alternatively, two auxiliary fuel tanks can be carried to aid self-deployment.

Low-rate production was scheduled to commence in 1996, but delays have pushed the date back to 2001.

Specification

Powerplant: two 1,002 kW (1,344 shp) LHTEC T800-LHT-801 turboshafts;

Dimensions: length (overall, rotor turning) 14.28 m (46 ft 10¼i n); height over tailplane 3.39 m (11 ft 1½ in); width over mainwheels 2.31 m (7 ft 7 in); main rotor diameter 11.90 m (39 ft ½ in)

Weights: empty 3,515 kg (7,749 lb); take-off weight, primary mission 4,587 kg (10,112 lb)

Performance (at 1,220 m (4,000 ft)): max level (dash) speed 328 km/h (204 mph); vertical rate of climb 360 m (1,182 ft)/min; endurance (standard fuel) 2 h 30 m

Armament: one 20 mm cannon with 320-500 rds, 10 Hellfire ATGMs, 14 Stinger AAMs and two 1,741 l (383 gal) fuel tanks

CASA C.101 Aviojet

The C.101 has been widely exported to South America

By far the most capable design to emerge from the Spanish aircraft industry, the C.101 was designed in conjunction with MBB of Germany and Northrop of the USA to meet a Spanish Air Force requirement for a new jet-powered primary trainer. First flight of this straightforward design took place on 27 June 1977, design features including a completely unswept wing and a tandem stepped cockpit.

The Spanish Air Force's trainer requirement was translated into an order for 88 examples of the first production model, the C.101EB-01, known in service as the E.25 Mirlo (Blackbird). Although up to seven hardpoints can be used (three beneath each wing and one centreline), the Spanish Air Force does not use its E.25s in a weapons training capacity. A customer that does use the aircraft in a more aggressive role is the Chilean Air Force, whose first acquisition comprised 14 C.101BB-02s (four from CASA and 10 assembled locally from CASA-supplied CKDs). Known in service as the T-36, these armed trainers entered service in late 1983, around the same time that the first A-36 Halcon took to the air. A-36 is the designation for 23 C.101CC-02s acquired by the Chilean Air Force, the C.101CC being a more powerful variant of the Aviojet family. Further export success came with orders from Honduras (four C.101BB-03s) and Jordan (16 C.101CC-04s).

Specification (C.101CC)

Powerplant: one 20.91 kN (4,700 lb st) Garrett TFE731-5-1J non-afterburning turbofan

Dimensions: length 12.50 m (41ft 0 in); height 4.25 m (13 ft 11¼ in); wing span 10.60 m (34 ft 9½ in)

Weights: empty, equipped 3,500 kg (7,716 lb); MTOW 6,300 kg (13,889 lb)

Performance: maximum level speed ('clean' at 6,095m (20,000 ft)) 806 km/h (501 mph); maximum rate of climb at sea level 1,494 m (4,900 ft)/min; service ceiling 12,800 m (42,000 ft)

Armament: optional twin Browning 12.7mm machine-gun pack with 220 rpg; up to 2,250 kg (4,960 lb) of BR250 bombs, AGM-65 Maverick ASMs, rocket launchers, AIM-9L/Magic AAMs and a centreline DEFA 553 30 mm cannon pod

Cessna A-37B/OA-37B Dragonfly

The A-37 proved to be a useful strike aircraft in Vietnam

Development of this diminutive light attack and reconnaissance aircraft can be traced back to a USAF requirement for a primary jet trainer in the early 1950s. Cessna's Model 318 was selected for development, and the first XT-37 prototype made its maiden flight on 12 October 1954. A total of 537 T-37A Tweets, the first production model were built, entering USAF service in 1957. They were subsequently upgraded to T-37B standard and joined by 447 new-build B-models, some of which were subsequently exported.

In the 1960s the USAF took delivery of another 252 T-37Cs, an export model equipped with two underwing pylons for light ordnance. An attack-dedicated development known as the A-37A Dragonfly was developed for the Vietnam war. The improved A-37B introduced a nose-mounted IFR capability and the more powerful J85-17A turbojet engine.

Some 577 A-37Bs were delivered up to the mid-1970s, many being supplied to smaller air forces, particularly in Central and South America, attracted to a small aircraft with such an impressive armament.

As for A-37Bs in USAF service, the end of the Vietnam war left them without a role. At least 130 were converted to OA-37Bs for use in the FAC (Forward Air Control) role, but by the early 1990s their time was up.

Specification (A-37B)

Powerplant: two 12.68 kN (2,850 lb st) General Electric J85-GE-17A non-afterburning turbojets

Dimensions: length (excluding IFR probe) 8.62 m (28 ft 3½ in); wing span (over tip tanks) 10.93 m (35 ft 10 in)

Weights: empty 2,817 kg (6,211 lb); maximum loaded 6,350 kg (14,000 lb)

Performance: maximum speed (MTOW at 4,875 m (16,000 ft)) 816 km/h (507 mph); initial rate of climb 2,130 m (6,990 ft)/min; service ceiling 12,730 m (41,765 ft)

Armament: one GAU-2B/A 7.62mm Minigun; up to 2,268 kg (5,000 lb) of bombs, rocket launchers and gun pods

Chengdu J-7/F-7

China's J-7 was developed from the Russian MiG-21

Thanks to the granting of a licence by the Soviet Union to allow manufacture of the MiG-21F-13 fighter and its Tumansky R-11F-300 turbjet engine, the People's Republic of China has been able to develop a family of capable single-seat fighters and close-support aircraft. Christened the Jianjiji-7 (Fighter aircraft 7), the first prototype took to the air on 17 January 1966. Shenyang went on to produce the early J-7s, but in June 1967 production by Chengdu commenced under the designation J-7I.

Development of the basic design continued in the mid-1970s with the J-7II, powered by a single Wopen WP7B afterburning turbojet engine. Examples for export (to Egypt and Iraq) were designated F-7B. The

F-7M Airguard, an upgraded version of the J-7II, represented an attempt to attract more overseas orders. Western avionics, more powerful engine, strengthened landing gear and two additional underwing stores points were its main selling points, and orders were indeed forthcoming. The Pakistan Air Force ordered a modified version of the F-7M, namely the F-7P Skybolt; 20 such aircraft were ordered, followed by 60 F-7MPs with improved cockpit layout and fin-mounted RWR.

Further development was based on the more advanced MiG-21MF and resulted in the J-7III. which has an all-weather day/night capability thanks to its JL-7 interception radar.

Specification (F-7M Airguard)

Powerplant: one 59.82 kN (13,488 lb st) Liyang Wopen-7B(BM) afterburning turbojet

Dimensions: length (including probe) 14.89 m (48 ft 10 in); height 4.10 m (13 ft 5½ in); wing span 7.15 m (23 ft 5¼ in)

Weights: empty 5,275 kg (11,629 lb); normal take-off 7,531 kg (16,603 lb)

Performance: maximum level speed ('clean' at high altitude) 2,175 km/h (1,350 mph); maximum rate of climb at sea level 10,800 m (35,433 ft)/min; service ceiling 18,200 m (59,700 ft)

Armament: two 30 mm cannon with 60 rpg; up to 1,000 kg (2,205 lb) of 50-500 kg (101-1,102 lb) bombs, PL-2/-2A/-5B/-7/Magic AAMs and 57 mm or 90 mm rocket pods

Dassault Atlantique 2

Dassault is supplying 42 Atlantique 2s to the French navy

Whereas other successful land-based ASW aircraft
have been developed from existing airliner designs
(for example, the BAe Nimrod from the de Havilland
Comet 4C and the Lockheed P-3 Orion from the
same company's Electra), the Dassault Atlantic twin-
turboprop design was an original. Built between 1964
and 1974, 87 were produced for service with the
navies of France, Germany, Italy and The Netherlands
(a small number were subsequently passed to the
Pakistan Navy).

In the late 1970s it was decided to develop an all-
new model, the Atlantique 2 (ATL 2). Two exisitng
Atlantic airframes were converted for use as
prototypes, the first one flying on 8 May 1981. The
new model's airframe was strengthened and received

extra anti-corrosion treatment. Other improvements were made to the on-board mission avionics. The Thompson-CSF Iguane radar offers track-while-scan coverage of up to 100 targets simultaneously. Antennae in the wing-tip and fin-top fairings feed data to the on-board ESM passive receiver, and at the rear of the aircraft the tell-tale 'stinger' houses a Crouzet MAD sensor. A small bay aft contains up to 78 TSM 8010/8020 sonobuoys.

So far, only the French Navy has put its faith in the Atlantique 2, with an order for 42 examples. Deliveries started in 1989 and are scheduled to continue until 2001.

Specification

Powerplant: two 4,549 kW (6,100 ehp) Rolls-Royce Tyne Mk 21 turboprops
Dimensions: length 33.63 m (110f t 4 in); height 10.89 m (35 ft 8¾i n); wing span (overwing-tip pods) 37.42 m (122 ft 9¼ in)
Weights: empty, equipped (standard mission) 25,700 kg (56,659 lb); MTOW 46,200 kg (101,850 lb)
Performance (at 45,000 kg (92,200 lb)): max level speed at sea level 592 km/h (368 mph); normal patrol speed at sea level 315 km/h (195mph); max rate of climb at sea level 884 m (2,900 ft)/min; service ceiling 9,145 m (30,000 ft)
Armament: (internal) up to eight depth charges, eight Mk46 homing torpedoes, two AM39 Exocet/AS37 Martel AShMs; (external) up to four ARMAT/Magic missiles, various equipment pods

Dassault Mirage IV

The Mirage IV was France's first nuclear bomber

In the early 1950s France decided to form its own nuclear deterrent force based on three delivery systems capable of carrying and launching nuclear weapons. One of these was to be a two-seat supersonic strategic bomber capable of long-range high-speed flights against targets in the Soviet Union. Naturally the French authorities turned to Dassault to answer such a requirement, and the company finally settled on a scaled-up version of the Mirage III, based on a twin-engined night-fighter design formulated in 1956.

The first prototype Mirage IVA took to the air on 17 June 1959, the Atar 09-powered aircraft achieving Mach 2 on its 33rd test flight. Three pre-production prototypes were subsequently built for use in the exhaustive test programme, these being larger in size and more representative of the production-standard Mirage IVA that began to enter service with the French Air Force during 1964. In all, 62 such aircraft were built and all had been delivered by 1968. Each carried a single 60 kT free-fall nuclear bomb.

Twelve Mirage IVAs were converted for long-range high-level recce missions. Of the remaining Mirage IVAs, 19 were upgraded to IVP (Penetration) standard during the 1980s, this including upgraded avionics, computers and cockpit displays, and a new weapon, the 30 kT ASMP nuclear cruise missile. Today, just 13 Mirage IVPs remain in service, three four-aircraft detachments operating from different locations for reasons of security.

Specification (Mirage IVP)

Powerplant: two 70.61 kN (15,873 lb st) SNECMA Atar 9K-50 afterburning turbojets

Dimensions: length 23.50 m (77 ft 1 in); height 5.65 m (18 ft 6½ in); wing span 11.85 m (38 ft 10½ in)

Weights: empty,equipped 14,500 kg (31,966 lb); MTOW 31,600 kg (69,666 lb)

Performance: max level speed ('clean' at 11,000 m (36,069 ft)) 2,338 km/h (1,453 mph); max speed at low level 1,350 km/h (840 mph); normal penetration speed (at 11,000 m (36,069 ft)) 1,913 km/h (1,189 mph); service ceiling 20,000 m (65,615 ft)

Armament: one 900 kg (1,984 lb) ASMP stand-off nuclear missile; conventional ordnance up to 7,200 kg (15,873 lb)

Dassault Mirage 5

The Mirage 5 is a ground attack version of the Mirage III

In 1966 Israel asked Dassault to create a simplified version of its Mirage IIIE strike/attack fighter optimized for the daylight ground-attack role. The type was evolved without radar, whose erstwhile volume was used for the avionics that were relocated from their position to the rear of the cockpit, which was now used for an additional 470 litres (103 Imp gal) of internal fuel tankage, and two outward-splayed hardpoints were added under the fuselage for the carriage of a heavier and more diverse warload. The first Mirage 5A flew on 19 May 1967, but the French government embargoed the delivery of the type to Israel and took the type for its own air force with the designation Mirage 5F.

This was only the start of a successful programme

that saw the delivery of 525 aircraft to 11 air forces, many of which still operate the type. The Mirage 5A is the single-seat fighter and ground-attack warplane, and variants include the Mirage 5D tandem two-seat trainer, the Mirage 5R reconnaissance type with a fan of five cameras, and the Mirage 50 final model with the uprated 70.60kN (15,873lb st) Atar 9K-50 turbojet and the electronic improvements retrofitted in most Mirage 5s.

Only Chile and Venezuela bought the Mirage 50, and the former has upgraded its aircraft, with the assistance of Israel Aircraft Industries, to Pantera standard with fixed canard foreplanes on the inlet trunks and further updated electronics.

Specification (Mirage 5A)

Powerplant: one 60.81 kN (13,670 lb st) SNECMA Atar 9C turbojet

Dimensions: length 15.55 m (51 ft ½ in); height 4.50 m (14 ft 9 in); wing span 8.22 m (26 ft 11½ in)

Weights: take-off ('clean') 9,600 kg (21,165 lb); MTOW 13,700 kg (30,203 lb)

Performance: max level speed at 12,000 m (39,370 ft) Mach 1.9 or 1912 km/h (1,188 mph); service ceiling 17,000 m (55,755f t)

Armament: two 30 mm DEFA 552A cannon with 125 rounds per gun; 4,000 kg (8,818 lb) of disposable stores, including AAMs, ASMs, bombs, rocket launchers, drop tanks and ECM pods, carried on seven external hardpoints.

Dassault Super Etendard

Super Etendards sank two British ships off the Falklands

In 1972 the French Navy decided it needed a replacement for its Vought F-8E(FN) Crusader fighter and instructed Dassault to create an updated version of its Etendard design. Work began in 1973, and the true prototype of the Super Etendard fighter and strike/attack warplane flew on 3 October 1975. The wing was adapted with additional high-lift devices, the avionics were updated and the Agave lightweight search radar fitted in a modified nose. Propulsion was improved by the Atar 8K-50 turbojet for marginally supersonic performance without afterburning,. It was fitted to carry the AM.39 Exocet anti-ship missile and R.550 Magic air-to-air missile.

The French Navy ordered 71 Super Etendards, and these were delivered between June 1978 and March 1983, although five aircraft were loaned to Iran in this period for use in its war with Iran, when several ships were sunk or damaged by Super Etendard-launched Exocets. Another 14 aircraft were delivered to the Argentine naval air arm, and although only five had been delivered by the time of the Falkands war of 1983, these succeeded in sinking two British ships while operating from a shore base. Some 50 of the Super Etendards have been fitted for the ASMP stand-off nuclear missile, and the type will be replaced early next century by the Dassault Rafale M.

Specification (Super Etendard)

Powerplant: one 49.03 kN (11,023 lb st) SNECMA Atar 8K-50 turbojet

Dimensions: length 14.31m (46 ft 11½ in); height 3.86 m (12 ft 8 in); wing span 9.60 m (31 ft 6 in)

Weights: take-off ('clean') 9,450 kg (20,835 lb); MTOW 12,000 kg (26,455 lb)

Performance: max level speed at 11,000 m (36,090 ft) Mach 1.3 or 1,380 km/h (857 mph); service ceiling more than 13,700 m (44,950 ft)

Armament: two DEFA 553 cannon with 125 rounds per gun; 2,100 kg (2,425 lb) of disposable stores, including nuclear weapons, ASMs, AAMs, bombs, rocket pods, drop tanks and ECM pods, carried on five external hardpoints.

Dassault Mirage III

A Mirage III of the Swiss Air Force

One of the most important commercial successes of
the French aero industry in the 1960s and 1970s, the
delta-winged Mirage III was conceived as a
lightweight interceptor but found its greatest success
as a multi-role fighter. After a number of
experimental developments that flew from 1956 with
the Atar 101G turbojet, the Mirage IIIA pre-
production type flew in May 1958 with the
considerably more powerful Atar 9B turbojet. This
paved the way for the two initial service models,
namely the Mirage IIIB tandem two-seat trainer
without radar and the Mirage IIIC single-seat
interceptor with Cyrano Ibis radar and provision for
the two cannon to be replaced by a rocket pack for

improved climb rate and ceiling. Subvariants of both models were exported to countries such as Israel, Lebanon, Switzerland and south Africa.

The definitive model was the Mirage IIIE optimized for the longer-range intruder and fighter-bomber roles with a longer fuselage, Cyrano II radar, Doppler and TACAN navigation systems, and provision for a wider assortment of disposable stores. This model was bought by France, and subvariants were exported to several countries. The type was also built under licence in Australia and Switzerland as the Mirage IIIO and Mirage IIIS respectively, the latter with Hughes radar and weapon system for use of the Hughes Falcon air-to-air missile.

Specification (Mirage IIIE)
Powerplant: one 94.17 kN (13,670lb st) SNECMA Atar 9C turbojet
Dimensions: length 15.03 m (49ft 3½in); height 4.50 m (14ft 9in); wing span 8.22 m (26 ft 11½ in)
Weights: take-off ('clean') 9,600 kg (21,165 lb); MTOW 13,700 kg (30,203 lb)
Performance: max level speed at 12,000 m (39,370ft) Mach 2.2 or 2,350 km/h (1,460 mph); service ceiling 17,000 m (55,755 ft)
Armament: two 30 mm DEFA 552A cannon with 125 rounds per gun; 4,000 kg (8,818 lb) of disposable stores, including nuclear weapons, ASMs, AAMs, bombs, rocket launchers, drop tanks and ECM pods, carried on five external hardpoints.

Dassault Mirage F1

Mirage F.1s were used by both sides in the 1991 Gulf War

Designed as successor to the delta-winged Mirage III/5 family, the Mirage F1 reverted to a conventional layout as many operators had complained about the earlier types' poor field performance and loss of energy in low-level manoeuvring flight. Dassault's design for the Mirage F1 was very clever, however, and yielded a major increase in internal fuel capacity despite a significant reduction in external area. The prototype flew in December 1966 and was designed in its initial Mirage F1C form as an all-weather interceptor with the Super 530 AAM. French orders were complemented by sales to Greece, Jordan,

Kuwait, Morocco, South Africa and Spain. South Africa wanted a radarless attack model, and this Mirage F1A was also sold to Ecuador and Libya. Just as the Mirage F1A was the analogue to the Mirage 5, the Mirage F1E was the counterpart to the Mirage IIIE for the strike and attack roles with upgraded electronics including radar. Sales of this type were made to Iraq, Jordan, Libya, Morocco, Qatar and Spain.

There are two tandem two-seat trainer models, namely the Mirage F1B and Mirage F1D based respectively on the Mirage F1C and Mirage F1E. Both models received useful orders, although the Mirage F1D received only one export order.

Specification (Mirage F1C)

Powerplant: one 70.21 kN (15,873 lb st) SNECMA Atar 9K-50 turbojet

Dimensions: length 15.23 m (49 ft 11½ in); height 4.50 m (14 ft 9 in); wing span (over tip missiles) 9.32 m (30 ft 10 in)

Weights: take-off ('clean') 10,900 kg (24,030 lb); MTOW 16,200 kg (35,714 lb)

Performance: max level speed at 12,000 m (39,370 ft) Mach 2.2 or 2350 km/h (1,460 mph); service ceiling 20,000 m (65,615 ft)

Armament: two 30 mm DEFA 553 cannon with 135 rounds per gun; 4,000 kg (8,818 lb) of disposable stores, including ASMs, AAMs, bombs, rocket launchers, drop tanks and ECM pods, carried on seven external hardpoints.

Dassault Mirage 2000C/2000-5

Like most Mirages, the 2000 is being widely exported

Chosen by the French Air Force in 1975 as its future main combat aircraft, the Mirage 2000 marked a return by Dassault to the familiar low-set delta wing configuration. Combined with an FBW flight control system and negative longitudinal stability to improve the overall handling characteristics, the result is an extremely agile and manoeuvrable fighter.

The first of five prototypes flew on 10 March 1978, and initial deliveries comprised 37 2000C interceptors. A major multi-role upgrade has been introduced in the form of the Mirage 2000-5. First flown on 24 October 1990, it can use the Super 530D or Sky Flash AAMs . Another option concerns the powerplant, with the SNECMA M88-P20 on offer for use later in the 1990s.

Export models include the Mirage 2000E (single-seat fighter), 2000ED (two-seat trainer) and 2000ER (single seat recce), supplied to Abu Dhabi, Egypt, Greece, India and Peru. Taiwan has ordered 60 M53-P2-powered Mirage 2000-5s.

Specification (2000C)

Powerplant: one 95.1 kN (21,385 lb st) SNECMA M53-P2 afterburning turbofan

Dimensions: length 14.36 m (47 ft 1¼ in); height 5.20 m (17 ft ¼in); wing span 9.13 m (29ft 11½in)

Weights: empty 7,500 kg (16,534lb); MTOW 17,000kg (37,480lb)

Performance: max level speed ('clean' at 11,000 m (36,069 ft)) +2,338 km/h (1,453mph); max rate of climb at sea level 17,060 m (55,971 ft)/min; service ceiling 16,460 m (54,000 f t)

Armament: two DEFA 554 30 mm cannon with 125rpg; up to 6,300 kg (13,889 lb) of ordnance, including BGL 1000 LGBs, BAP-100 anti-runway bombs, 250 kg (551 lb) retarded bombs, Belouga cluster bombs, AS30L/ARMAT ARMs, 68 mm or 10 mm rockets, Super 530D/530F AAMs, Magic/Magic 2 AAMs, MICA AAMs, three auxiliary fuel tanks

Dassault Mirage 2000N/D/S

The Mirage 2000N was designed as a new nuclear bomber

In 1979, with prototypes of the Mirage 2000 undertaking flight-test work, Dassault was awarded a contract to build two prototypes of what was designated the Mirage 2000P (Penetration), later redesignated Mirage 2000N (Nuclear). Based on the Mirage 2000B two-seat trainer, the programme's aim was to develop a low-altitiude replacement for the French Air Force's fleet of nuclear-armed Mirage IVPs.

First flown on 3 February 1983, the Mirage 2000N incorporates a heavily revised avionics suite and strengthened fuselage for its low-altitude role. The terrain-following Antilope 5 radar enables automatic flight down to 61m (200 ft) at speeds below 1,112km/h (691mph).

The conventional weapons-carrying abilities of the

2000N-K2 have been further developed in the Mirage 2000D. First flown on 19 February 1991, the 2000D cannot carry the ASMP missile but has been built in two configurations to date: R1N1L (LGWs and Magic AAMs only) and R1 (full range of conventional weapons). The planned R2 configuration, to be introduced during the second half of the 1990s, will confer a fully integrated self-defence suite and compatibility with the Apache stand-off weapons dispenser. An export model, the Mirage 2000S (Strike), is also available.

Specification (2000N)

Powerplant: one 95.1 kN (21,385 lb st) SNECMA M53-P2 afterburning turbofan

Dimensions: length 14.55m (47 ft 9 in); height 5.15 m (16 ft 10¾ in); wing span 9.13 m (29 ft 11½in)

Weights: empty 7,600 kg (16,755 lb); MTOW 17,000 kg (37,480 lb)

Performance: max level speed ('clean' at 11,000 m (36,069 ft)) +2,338 km/h (1,453 mph); max rate of climb at sea level 17,060 m (55,971ft)/min; service ceiling 16,460 m (54,000 ft)

Armament: one 900 kg (1,984 lb) ASMP tactical nuclear missile; up to 6,300 kg (13,889 lb) of ordnance including BGL 1000 LGBs, BAP-100/Durandal anti-runway bombs, 250 kg (551 lb) retarded bombs, Belouga cluster bombs, APACHE munitions dispensers, AS30L/ARMAT ARMs, AM39 Exocet AShMs, 68 mm or 10 mm rockets, CC630 twin 30mm gun pods, three auxiliary fuel tanks

Dassault Rafale

Rafales will replace the French navy's ageing Vought F-8s

Although France was part of the EFA international project, work was simultaneously underway on an advanced combat aircraft. France's withdrawal from the EFA project in 1985 lent further impetus to the ACX, or Rafale A, which first flew on 4 July 1986, The single-seat Rafale C flew for the first time on 19 May 1991, followed by the navalized Rafale M prototype on 12 December 1991.

The Rafale C is the single-seat multi-role combat version ordered by the French Air Force. Fourteen external hardpoints can carry up to 8,000 kg (13,228lb) of ordnance including the ASMP nuclear stand-off missile. The Rafale M is 610kg (1,345lb) heavier. Navalization includes an arrestor hook and extended nosewheel oleo and one less centreline stores point.

Rafale B was originally intended as a two-seat dual control trainer for the French Air Force, but is now being developed as a fully operational variant in its own right.

Current plans call for 94 Cs, 78 Ms and 140 Bs to be built. Deliveries are scheduled to begin in 1998 and run until 2009, with the first 20 S01-standard Rafale Ms replacing veteran F-8P Crusaders followed by SO2-standard aircraft replacing Super Etendards.

Specification (Rafale M)

Powerplant: two 87 kN (19,558 lb) SNECMA M88-3 afterburning turbofans

Dimensions: length 15.30 m (50 ft 2½ in); height 5.34 m (17 ft 6¼ in); wing span (over AAMs) 10.90 m (35 ft 9¼ in)

Weights (estimated): empty, equipped 9,670 kg (21,319 lb); MTOW 21,500 kg (47,399 lb)

Performance (estimated): max level speed ('clean' at 11,000 m (36,069 ft)) 2,125 km/h (1,321mph); max speed at low level 1,390 km/h (864 mph)

Armament: one GIAT DEFA 791B 30 mm cannon; up to 6,000 kg (13,228 lb) of ordnance including one 900 kg (1,984 lb) tactical nuclear missile, stand-off munitions dispensers, bombs, laser designator pod or auxiliary fuel tanks

Dassault-Dornier Alpha Jet

Ex-German Alpha Jets are now operated by Portugal

The result of a Franco-German collaboration programme to develop a twin-engined jet trainer and light attack aircraft, the Alpha Jet entered production in France (as the Alpha Jet E) in 1977, and a year later in Germany (as the Alpha Jet A). Both models share the same basic features such as a shoulder-mounted wing, underfuselage landing gear and stepped tandem cockpits, but their different roles led to appreciable differences in systems fits. For the Luftwaffe, the Alpha Jet A was to replace its Fiat G.91R/3s in the light attack/close-support role. Consequently, German aircraft featured an advanced nav/attack system including HUD, Doppler navigation radar, and twin-gyro INS. The underfuselage 30mm gun was replaced by a 27mm Mauser cannon; alternatively, a Super Cyclope recce pod could be carried on the centreline station.

The Alpha Jet has found considerable export success. The Portuguese Air Force received 50 ex-Luftwaffe aircraft in 1994. The Egyptian Air Force ordered 30 including 26 licence-built trainers designated Alpha Jet MS1. These were followed by 15 MS2s, optimized for training/light attack. The MS2 led in turn to the better equipped Alpha Jet 3 Advanced Training System (also known as the Lancier), featuring two MFDs in each cockpit.

Specification (Alpha Jet E)

Powerplant: two 13.24 kN (2,976 lb st) SNECMA/Turbomeca Larzac 04-C6 turbofans

Dimensions: length 11.75 m (38 ft 6½ in); height 4.19 m (13 ft 9 in); wing span 9.11 m (29 ft 10¾ in)

Weights: empty, equipped 3,345 kg (7,374 lb); MTOW 8,000 kg (17,637 lb)

Performance: max level speed ('clean' at sea level) 1,000 km/h (621 mph); max rate of climb at sea level 3,660 m (12,008 ft)/min; service ceiling 14,630 m (48,000 ft)

Armament: one Mauser 27 mm or DEFA 30 mm ventral cannon pod with 125rds; up to 2,500 kg (5,511 lb) of bombs, rockets, missiles, recce/ECM pods and auxiliary fuel tanks

Eurocopter BO 105

The BO105 was the first German anti-tank helicopter

Originally conceived as a light helicopter for the commercial market, the BO 105 had obvious military potential. First flown on 16 February 1967, the BO 105's most ardent military fan has been the German Army, with orders for 100 BO 105 scouts and a further 212 configured for anti-armour operations. Based on the familiar pod and boom configuration, the BO 105 differs from many of its similar-sized contemporaries in being powered by two turboshaft engines, these being housed high in the rear of the cabin.

The German Army decided in 1974 to acquire a military version of the commercial BO 105CB.

Designated BO 105M by the company, it is known in military service as the VBH and is used for scouting, liaison and communications duties. The survivors operate alongside the PAH-1 (BO 105P), some 200 of which remain in service.

Development of the PAH-1 began in 1977, its primary anti-armour weapon being the Euromissile HOT ATGM (anti-tank guided missile), used in conjunction with an SFIM APX M397 roof-mounted direct-view optical sight. The PAH-1 force has been upgraded to PAH-1-1A standard, improvments including new rotor blades and engine air intakes, as well as the ability to fire up to six HOT 2 ATGMs.

Specification (PAH-1)

Powerplant: two 313 kW (420 shp) Allison 250-C20B turboshafts
Dimensions: length (rotors turning) 11.86 m (38 ft 11 in), fuselage pod 4.55 m (14 ft 11 in); height 3.00m (9 ft 10 in); main rotor diameter 9.84 m (32 ft 3½ in)
Weights: empty 1,458 kg (3,214 lb); MTOW 2,380 kg (5,247 lb)
Performance: never-exceed speed at sea level 242 km/h (150 mph); max rate of climb at sea level 441m (1,457 ft)/min; max operating altitude at 2,500 kg (5,511 lb) 3,050 m (10,000 ft)
Armament: up to six HOT or eight TOW 2 ATGMs, 70 mm rocket pods, machine-gun pods and 20 mm cannon

Eurocopter SA 330 Puma/AS 532 Cougar

Originally a transport, the Puma now has many roles

Designed in the mid-1960s to meet a French Army requirement for a new medium-lift transport helicopter, the Aerospatiale Puma was subsequently produced in cooperation with the United Kingdom to meet the RAF's need for a tactical transport helicopter. The first flight took place on 15 April 1965, and close to 700 Pumas had been built by the time production ended in 1984. Of these, the vast majority were unarmed transports, but many of the 100 or so built under licence in Romania can carry anti-tank missiles and machine-gun pods.

The South African Air Force had two Puma transports converted to XTP-1s to act as development platforms for the CSH-2 Rooivalk attack helicopter.

The Puma gunship that has evolved is fitted with a nose-mounted HSOS (helicopter stablised optronic sight) containing FLIR, TV, laser rangefinder and autotrack. It is armed with eight ZT3 127mm ATGMs or up to 72 68mm HR-68 rockets plus a 20mm cannon beneath the fuselage.

Aerospatiale replaced the Puma with the Super Puma, first flown on 13 August 1978, military variants of which were christened Cougar I in 1990. The unarmed AS 532UC and UL have been offered in armed configuration as the AS 532AC and AL respectively; the AS 532SC is a navalized model that carries a pair of side-mounted AM39 Exocet missiles. The AS 532UL Horizon battlefield surveillance helicopter is being developed for the French Army.

Specification (AS 532U2)

Powerplant: two 1,573 kW (2,109 shp) Turbomeca Makila 1A2 turboshafts

Dimensions: length (main rotor folded) 16.79 m (55 ft ½ in); height (to top of rotor head) 4.60 m (15 ft 1 in); main rotor diameter 16.20 m (53 ft 1½in); width 3.38 m (11 ft 1 in)

Weights: empty 4,760 kg (10,493 lb); MTOW 9,750 kg (21,495 lb); max slung load 4,500 kg (9,920 lb)

Performance: never-exceed speed 327 km/h (203 mph); rate of climb at sea level (at 130 km/h (81 mph)) 384 m (1,260 ft)/min; service ceiling 4,100 m (13,451 ft); endurance (standard fuel) at 130 km/h (81mph) 4 h 12 min

Eurocopter SA 341/342 Gazelle

Gazelles are now being fitted with anti-tank missiles

Designed in the mid-1960s to replace French Army Alouette IIs, Aerospatiale's Project X.300 (soon redesignated SA 340) combined its predecessor's Astazou II powerplant and transmission with a new three-blade rigid main rotor, an enclosed fuselage and the revolutionary fan-in-the-fin 'fenestron'. Subsequently christened Gazelle, it became one of three helicopters to be manufactured as part of the Anglo-French helicopter agreement of 1967.

The vast majority of SA 341s built were unarmed, but some 40 of the 157 SA 341Fs for the French Army have been redesignated SA 341M and carry four HOT ATGMs. Another 60 or so SA 341Fs can carry a 20 mm cannon. In Mostar, Yugoslavia (now part of Bosnia-Herzegovina) Soko built SA341Hs fitted with AT-3 anti-tank missiles until the civil war stopped production in 1992.

Improvement of the SA 341 concentrated on an improved powerplant. First was the unarmed SA 342K, followed by the armed SA 342L. The third sub-variant is the SA 342L2 ATAM, this designation covering 30 French Army Gazelles armed with four Mistral AAMs and a Sextant 200 sight. Finally, there is the Astazou XIVM-powered SA 342M, also in service with the French Army, 70 of the 157 examples acquired currently being modified to enable night-firing of HOT missiles.

Specification (SA 342L2)

Powerplant: one 640 kW (858 shp) Turbomeca Astazou XIV M2 turboshaft

Dimensions: length (rotors turning) 11.97 m (39 ft 3 in); height (to top of fin) 3.19 m (10 ft 5½ in); main rotor diameter 10.50 m (34 ft 5½ in)

Weights: empty 920 kg (2,208 lb); MTOW 2,100 kg (4,630 lb)

Performance: never-exceed speed at sea level 310 km/h (193mph); max cruising speed at sea level 260 km/h (161mph); max rate of climb at sea level 540 m (1,770 ft)/min; service ceiling 5,000 m (16,405 ft)

Armament: up to 700 kg (1,540 lb) of ordnance including four Mistral AAMs, rocket pods, gun pods

Eurocopter AS 550/555 Fennec

The Fennec serves as a utility helicopter or gunship

With over 2,000 examples built to date, the Ecureuil/Astar/Twin-Star family of light general-purpose helicopters have proved highly successful. With a crew of two and seating for four in a cabin that can be configured to meet various operational requirements, it is hardly surprising that armed variants have been developed of this versatile helicopter which first flew in single-engined prototype form on 27 June 1974. These militarized variants are known collectively as the AS 550/555 Fennec.

The AS 550 is a militarized version of the single-engined AS 350B2 Ecureuil. Powered by an Arriel

1D1 turboshaft, its sub-variants include the armed AS 550A2, anti-tank AS 550C2 and armed navalized AS 550S2. Twelve C2s have been acquired by the Royal Danish Army, and the Brazilian Army operates locally-assembled A2s (known as HA-1 Esquilos).

The AS 555 is the military version of the twin-engined AS 355EC2 Twin-Star, first flown on 28 September 1979. Like the single-engined AS 550, the AS 555 has armed sub-variants. The armed AS 555AN can carry a centreline 20mm gun, as well as Mistral AAMs on side-mounted pylons. The AS 555CN is a missile-armed model, and the navalized AS 555N can carry a homing torpedo or cannon/rockets for ASW. AS 550U2 and AS 555U2 utility variants of both models have also been built by Helibras of Brazil as the CH-50/-55 (Brazilian Air Force) and UH-12/-12B Esquilo (Brazilian Navy).

Specification (AS 550)

Powerplant: one 546 kW (641shp) Turbomeca Arriel 1D1 turboshaft

Dimensions: length (rotors turning) 12.94 m (42 ft 5½ in); height 3.34 m (10 ft 11½ in); width 2.53 m (8 ft 3¾ in); main rotor diameter 10.69 m (35 ft ¾ in)

Weights: empty 1,220 kg (2,689lb); MTOW 2,250 kg (4,960lb)

Performance (at 2,200 kg (4,850lb)): never-exceed speed at sea level 287 km/h (178mph).

Eurocopter AS 565 Panther

The multi-role Panther has been ordered by Brazil

First flown in protoytpe (AS 365M) form on 29 February 1984, the AS 565 Panther is based on the highly successful twin-engined AS 365N2 Dauphin 2. Today, armed and unarmed variants comprise the AS 565 Panther family, and can best be divided into those configured for Army/Air Force or Naval applications.

In the first category, the AS 565UA is an unarmed high-speed assault transport, its cabin being used to carry 8-10 troops. Further development led to the AS 565AA, an armed variant with fuselage outriggers for two rocket packs or gun pods, or four AAMs. The Brazilian Army was the launch customer for this model with an order for 36. Armed with HOT

ATGMs, the AS 565CA dedicated anti-tank variant can pack a bigger punch. A roof-mounted sight is used in conjunction with the missiles.

Like their Army/Air Force counterparts, navalized Panthers include unarmed (AS 565MA) and armed (AS 565SA) variants. The former is used for SAR and sea surveillance and was launched with an order from the Royal Saudi Navy for four such helicopters. The French Navy plans to acquire up to 40. Saudi Arabia was also the launch customer for the armed AS 565SA, with an order for 20 configured for anti-ship duties. Panthers assigned to anti-ship work feature a chin-mounted, roll-stabilized Agrion 15 radar and up to four AS.15TT AShMs; those for ASW have a MAD or sonar and two homing torpedoes.

Specification (AS 565SA)

Powerplant: two 558 kW (749 shp) Turbomeca Arriel IM1 turboshafts

Dimensions: length (fuselage) 12.11m (39 ft 8¾ in); height 3.99 m (13 ft 1 in); width over missiles 4.20 m (13 ft 9½ in); main rotor diameter 11.94 m (39 ft 2 in)

Weights: empty 2,262 kg (4,987 lb); MTOW 4,250 kg (9,370 lb)

Performance (at 4,000 kg (8,818 lb)): never-exceed speed 296 km/h (184 mph); max cruising speed at sea level 274 km/h (170 mph); max rate of climb at sea level 420 m (1,380 ft)/min

Armament: four ASMs or two homing torpedoes

Eurocopter Tiger/Tigre

The Tiger carries a formidable armament of missiles

A Franco-German agreement in 1984 to develop a new anti-tank helicopter led to approval in 1989 for full-scale development of a design known as both the Tiger (Germany) and Tigre (France). The first example (PT1) took to the air on 27 April 1991 and has since been fitted with aerodynamic mock-ups of the mast-mounted and roof-mounted sights planned for production helicopters.

As with other modern anti-tank helicopters, the Tiger/Tigre adopts a tandem two-seat configuration with the CP/G in the front cockpit and the pilot in the raised rear cockpit. Tigre HAP (formerly Gerfaut) is an escort/fire support model for the French Army. Systems include a roof-mounted TV sight, FLIR, direct-view optics and a laser rangefinder. Primary armament will comprise a 30mm undernose cannon and armour-piercing rockets and AAMs on up to four stub-wing stations. Deliveries to the French Army are scheduled to begin in 1999, which will coincide with the first deliveries of the French Army's second Tigre model, the HAC. This is a dedicated anti-tank model armed with up to eight ATGMs and four AAMs. It will have a prominent mast-mounted sight, as per the German Army's UHU model.

Specification

Powerplant: two 958 kW (1,258 shp) MTU/Rolls-Royce/Turbomeca MTR 390 turboshafts

Dimensions: length 14.00 m (45 ft 1¼ in); height (to top of tail rotor disc) 4.32 m (14 ft 2 in); main rotor diameter 13.00 m (42 ft 7¾ in)

Weights: empty 3,300 kg (7,275 lb); mission take-off weight 5,300-5,800 kg (11,685-12,787 lb); maximum overload take-off weight 6,000 kg (13,227 lb)

Performance (estimated): cruising speed 250-280 km/h (155-174 mph); max rate of climb at sea level +600 m (1,970ft)/min; endurance (with 20 min fuel reserves) 2 h 50 min

Armament: (HAP) one GIAT AM-30781 30 mm cannon with 150-450 rds, four Mistral AAMs, two 22-round 68 mm rocket pods, two 12-round rocket pods

Eurofighter 2000

The Eurofighter is still delayed by political problems

Flown for the first time on 29 March 1994 amid much political controversy and recrimination, Eurofighter 2000 (formerly EFA) is a single-seat air defence/air superiority fighter with a secondary ground attack capability. A two-seat trainer derivative will also be combat-capable.

Much of the criticism surrounding the aircraft has come from Germany, one of five nations (Spain, Italy, France and the UK being the others) who produced a requirement in 1983 for a common fighter design for deployment in the late 1990s. The French left the project in 1985 (favouring their own Rafale), and the now four-nation programme continued until 1992. Then Germany insisted that cheaper design proposals be studied. No suitable alternative was found, and so

114

the project was relaunched in late 1992.

Seven development aircraft (DA1-DA7) will conduct the test programme for what is a low-wing tailless delta aircraft relying heavily on CFCs in its structure. All-moving foreplanes and a quadruplex FBW flight control system enhance its agility. Main sensors are the ECR90 multi-mode Doppler radar and Eurofirst PIRATE, the latter housed in a fairing close to the port side of the windscreen.

As of late 1994, revised orders (including two-seaters) stand at around 600 (UK 250, Germany 120-140, Italy 130, Spain 87), all but the UK having reduced their requirements.

Specification

Powerplant: two 90 kN (20,250 lb st) Eurojet EJ200 afterburning turbofans

Dimensions: length 14.50 m (47 ft 7 in); height 6.40 m (21 ft 0in); wing span 10.50 m (34 ft 5½ in)

Weights: empty 9,750 kg (21,495 lb); MTOW 21,000 kg (46,297 lb)

Performance: max level speed ('clean' at 11,000 m (36,069 ft)) 2,125 km/h (1,321 mph); take-off/landing distance (full internal fuel, two AIM-120 and two dogfight missiles) 500 m (1,640 ft)

Armament: one Mauser Mk27 27mm cannon; up to 6,500kg (14,330lb) of ordnance including AIM-120 AMRAAMs, Aspide AAMs, short-range AAMs, various air-to-surface weapons, three auxiliary fuel tanks

115

Fairchild A/OA-10A Thunderbolt II

A-10s destroyed hundreds of enemy tanks in the Gulf War

Nicknamed the 'Warthog', the A-10A evolved as a result of USAF experience in Vietnam which highlighted the need for a new close support aircraft with anti-tank capability. Fairchild's design sacrificed sleekness for survivability and operational effectiveness; the low-set large-area wing offers extremely good low-speed manoeuvrability over the battlefield, while the two turbofans are housed in separate external pods positioned towards the rear of the fuselage. Armour protection is the key to the pilot's survival: he is seated in a titanium 'bathtub', protection extending to the nearby ammunition tank for the A-10A's most fearsome weapon, its 30mm seven-barrel rotary cannon (capable of firing depleted uranium shells at 2,100 or 4,200rpm). Eight underwing and three underfuselage hardpoints can carry up to 7,258 kg (16,000lb) of ordnance, the most potent weapon being the AGM-65 Maverick ASM. Beneath the starboard forward fuselage is a Pave Penny seeker that allows the pilot to spot targets 'painted' by laser. Avionics remain very simple and straightforward, although 90 aircraft have been upgraded with radar altimeter, GPWS, autopilot, improved bomb-sight and an air-to-air capability for the 30mm cannon.

A total of 721 A-10As entered operational service with the USAF, making a spectacular contribution to the Gulf War of 1991. A small number have been redesignated OA-10As for use in the FAC role.

Specification (A-10A)

Powerplant: two 40.32 kN (9,065 lb st) General Electric TF34-GE-100 non-afterburning turbofans

Dimensions: length 16.26 m (53 ft 4 in); height 4.47 m (14 ft 8in); wing span 17.53 m (57 ft 6 in)

Weights: empty, equipped 11,321 kg (24,959 lb); MTOW 22,680 kg (50,000 lb)

Performance: max level speed at sea level 706 km/h (439 mph); max rate of climb at sea level 1,828 m (6,000 ft)/min

Armament: one GAU-8/A 30 mm cannon with 1,174rds; up to 7,258 kg (16,000 lb) of ordnance including AGM-65 Maverick ASMs, LGBs, free-fall bombs, ECM pods, AIM-9 AAMs, auxiliary fuel tanks

FMA IA 58 Pucara

Pucaras flew from the occupied Falklands in 1982

Named after a form of South American stone hill fortress, the Pucara's origins can be traced back to the mid-1960s when Argentina's Fabrica Militar de Aviones (Military Aircraft Factory) was requested to develop a new combat aircraft capable of performing COIN, CAS and reconnaissance missions. The first flight of the prototype AX-2 Delfin, powered by a pair of Garrett TPE331-U-303 turboprops, took place on 20 August 1969. Subsequent prototypes were re-engined with French Turbom Èca Astazou XVIG turboprops.

The Pucara was designed to operate from rough-field and unprepared sites with the minimum of ground support - a point it proved to good effect during the Falklands War of 1982. Operations are possible by night, but not in adverse weather conditions, and weapons aiming is achieved visually by the pilot making full use of the excellent forward visibility over the Pucara's downward-sloping nose.

The production-standard IA 58A first flew on 8 November 1974, with deliveries to the Argentinian Air Force commencing just over a year later. Improvements led to the IA 58B model, the main upgrades being improved avionics and the additon of two 30mm cannon in place of the 20mm weapons. However, overall production figures have been modest at best, with exports to Uruguay, Sri Lanka and Colombia accounting for less than 20 aircraft in total.

Specification (IA 58A)

Powerplant: two 988 shp Turbomeca Astazou XVIG turboprops
Dimensions: length 15.25 m (46 ft 9 in); height 5.36 m (17 ft 7in); wing span 14.50 m (47 ft 7 in)
Weights: empty, equipped 4,037 kg (8,900 lb); MTOW 6,800 kg (14,991 lb)
Performance: max speed at 3,000 m (9,840 ft) 500 km/h (311 mph); economic cruising speed 430 km/h (267 mph); service ceiling 9,700 m (31,825 ft)
Armament: two Hispano HS804 20 mm cannon each with 270 rpg, four FN Browning 7.62 mm cannon with 900 rpg; up to 1,500 kg (3,307 lb) of free-fall bombs, napalm tanks, 70 mm (2.75 in) rockets, cannon pods, two auxiliary fuel tanks

General Dynamics F-111

F-111s led the US precision bombing attack on Libya

Nicknamed 'Aardvark' because of its long, slightly upturned nose, the F-111 evolved in response to a joint services requirement in the 1960s for a long-range interceptor (US Navy) and deep-strike interdictor (USAF). Flown in prototype form on 21 December 1964, the F-111 was highly innovative for its time, it being the first aircraft to feature a production variable-geometry wing and a fully enclosed detachable escape module for the crew.

Over Vietnam, the USAF's first deployment of F-111As led to an unacceptably high loss rate and the US Navy's F-111B was cancelled after it was found to be too heavy for carrierborne operations. In total, 144

F-111As ere built and 76 FB-111A strategic bombers. Twenty-four ex-USAF FB-111As have been acquired by the RAAF, joining the survivors of 28 F-111As exported to the RAAF as F-111Cs. Next came the F-111D which proved to be troublesome to maintain.

Ninety-six F-111Es were built, similar to the F-111A but with enlarged engine air intakes and more powerful engines. A small number remain in service, but the majority of active USAF 'Aardvarks' are F-111Fs, the final operational variant, of which 106 were built. Simpler avionics, more reliable engines and the ability to use laser-guided weapons make this by far the most effective member of the F-111 family.

Specification (F-111F)

Powerplant: two 111.65 kN (25,100 lb st) Pratt & Whitney TF30-P-100 afterburning turbofans

Dimensions: length 22.40 m (73 ft 6 in); height 5.22 m (17 ft 1½in); wing span (fully swept) 9.74 m (31 ft 11½ in), (fully spread) 19.20 m (63 ft 0 in)

Weights: empty, equipped 21,537 kg (47,481 lb); MTOW 45,360 kg (100,000 lb)

Performance: max level speed ('clean' at 11,000 m (36,069 ft)) 2,660 km/h (1,653 mph); cruising speed, penetration 919 km/h (571 mph)

Armament: up to 14,228 kg (31,500 lb) of ordnance including B43/B61 nuclear bombs, GBU-10/-12/-24 LGBs, GBU-15/28 EO-guided bombs, free-fall bombs, cluster bombs, Durandal anti-runway bombs, AIM-9P-3 self-defence AAMs, auxiliary fuel tanks

General Dynamics F-16 Fighting Falcon

The F-16 is one of the most widely exported US fighters

One of the world's most important and versatile
warplanes, the Fighting Falcon, was designed by
General Dynamics from the early 1970s but is now a
product of Lockheed since its purchase of General
Dynamics' fighter division. Designed to meet the US
Air Force's requirement for a light-weight fighter, the
YF-16 first flew on 2 February 1974 and emerged as
victor from a competitive evaluation against the
Northrop YF-17. The key elements in the design are a
very high power/weight ratio for good performance,
especially in climbing and turning flight, a fly-by-wire
control system so that maximum agility can be
extracted from a layout of relaxed static stability, a

semi-reclining pilot's seat under a clear-view canopy offering unexcelled fields of vision, and advanced yet flexible avionics for long-range target acquisition and accurate weapon delivery. The Fighting Falcon was conceived as an air combat fighter, but has matured as a truly exceptional multi-role fighter.

Operators of the F-16 include Bahrain, Belgium, Denmark, Egypt, Greece, Indonesia, Israel, the Netherlands, Norway, Pakistan, Portugal, Singapore, South Korea, Thailand, Turkey, the USA and Venezuela. Production is scheduled to end later in this decade, but further models will certainly emerge.

Specification (F-16C Fighting Falcon)

Powerplant: one 129.0 kN (29,000 lb st) General Electric F110-GE-129 or Pratt & Whitney F100-P-229 turbofan

Dimensions: length 15.03 m (49 ft 4 in); height 5.09 m (16 ft 8½in); wing span (over tip missile launchers) 9.45 m (31 ft 0 in)

Weights: take-off ('clean') 9,791 kg (21,585 lb); MTOW 19,187 kg (42,300 lb)

Performance: max level speed at 12,190 m (40,000 ft) more than Mach 2 or 2,124 km/h (1,320 mph); service ceiling more than 15,240 m (50,000 ft)

Armament: one 20 mm M61A1 Vulcan three-barrel cannon with 515 rounds; 9276 kg (20,450 lb) of disposable stores, including nuclear weapons, ASMs, AAMs, anti-radar missiles, anti-ship missiles, free-fall or guided bombs, cluster bombers, dispenser weapons, rocket launchers, napalm tanks, drop tanks and ECM pods, carried on nine external hardpoints.

Grumman A-6E Intruder

A-6s served the US Navy well in Vietnam and the Gulf war

The Korean War highlighted the US Navy/Marine Corps' need for a new all-weather long-range carrierborne attack aircraft capable of operating by day or night. A subsequent competition was won in 1957 by Grumman, the first of six YA2F-1 prototypes flying for the first time on 19 April 1960. These were followed by 482 production A-6As delivered to the US Navy from early 1963. They played a major role in the Vietnam war, flying some 35,000 sorties by 1973. An unarmed variant, the KA-6D, was developed as the US Navy's principal carrier-based tanker. The A-6E is typical of the whole A-6 family, with long-span wings, almost full-span flaps and distinctive wing-tip split airbrakes. The bulbous nose radome remains, but now it houses a single AN/APQ-148 multi-mode radar capable of track-while-scan and ground-mapping terrain-avoidance. Beneath the radome is a stabilized chin turret containing IIR, laser spot tracker and laser ranger/designator.

A proposed A-6F would have featured a new airframe and F404 powerplants but it was rejected in favour of the A-12 Avenger II. Cancellation of the latter in 1991 has resulted in further upgrading of the surviving A-6Es, but a rewinging programme was abandoned, and the remaining A-6Es are now due to be retired by the end of the century.

Specification (A-6E)

Powerplant: two 41.37 kN (9,300 lb st) Pratt & Whitney J52-P-8B non-afterburning turbojets

Dimensions: length 16.69 m (54 ft 9 in); height 4.93 m (16 ft 2 in); wing span 16.15 m (53 ft 0 in)

Weights: empty 12,525 kg (27,613 lb); MTOW (catapult launch) 26,580 kg (58,600 lb)

Performance: max level speed at sea level 1,037 km/h (644 mph); max rate of climb at sea level 2,323 m (7,620 ft)/min; service ceiling 12,925 m (42,400 ft)

Armament: up to 8,165 kg (18,000 lb) of ordnance including free-fall/retarded bombs, AGM-65 Maverick ASMs, AGM-84 Harpoon AShMs, AGM-84E SLAMs, AGM-88 HARMs, four auxiliary fuel tanks

Grumman F-14A Tomcat

The F-14 is the world's greatest long range interceptor

Suspecting that the F-111B version of the General Dynamics F-111 land-based strike fighter it was developing for the US Navy would be too heavy for its intended carrierborne fleet defence task, Grumman was already preparing an alternative fighter when the F-111B was cancelled in the later 1960s. Grumman exploited the experience it had gained from the F-111B in the variable-geometry wing planform, and also inherited, from the F-111B, the Pratt & Whitney TF30 afterburning turbofan and the Hughes weapon system, based on the AWG-9 radar and fire-control system together with the AIM-54 long-range AAM.

The G-303 design was preferred over three other contenders, and the first F-14A flew on 21 December 1970. Despite the loss of the prototype as a result of a

126

hydraulic failure, development was relatively trouble-free, and the first of 478 aircraft for the US Navy was delivered in October 1972 for an initial operational capability in September 1974. Another 79 aircraft were delivered to Iran.

The Tomcat is, without doubt, still the most potent long-range interceptor in service anywhere in the world, its provision for carriage of up to six Phoenix AAMs bestowing the capability for the simultaneous destruction of six targets at ranges in excess of 160km (100 miles). Closer engagements are also possible with the AIM-7 Sparrow AAM, AIM-9 Sidewinder AAM and M61 Vulcan cannon.

Specification (F-14A Tomcat)

Powerplant: two 92.97 kN (20,900 lb st) Pratt & Whitney TF30-P-412A/414A turbofans

Dimensions: length 19.10 m (62 ft 8 n); height 4.88 m (16 ft 0 in); wing span 19.55 m (64 ft 1½ in) spread and 11.65 m (38 ft 2½ n) swept

Weights: take-off ('clean') 26,553 kg (58,539 lb); MTOW 33,724 kg (74,379 lb)

Performance: max level speed at 10,975 m (36,000 ft) Mach 2.37 or 2517 km/h (1,564 mph); service ceiling more than 17,070 m (56,000 ft)

Armament: one 20 mm M61A1 Vulcan six-barrel cannon with 675 rounds; 6,577 kg (14,500 lb) of disposable stores, including six AIM-54 Phoenix, or six AIM-7 Sparrow

Grumman F-14B/D Tomcat

The F-14B/D takes the Tomcat design into the next century

After the problems with the TF30 powerplant Grumman produced a single Tomcat powered by a pair of F401-PW-400 turbofans. It became the test-bed for the General Electric F101-X Christened the F-14B Super Tomcat, flight trials began in July 1981 and it was decided to develop the engine to power proposed second-generation production Tomcats known as the F-14A+ (interim converted F-14As) and F-14D (new-build aircraft). The F-14+ became the F-14(Plus), then the F-14B, this last designation applying to 32 rebuilt F-14As and 38 new-build aircraft. Full-scale development began in 1984, and the first flight of a production F-14B took place on

14 November 1987. F-14Bs now equip two US Navy squadrons, with the possibility that more converted F-14As might follow.

As for the new-build F-14D (Super Tomcat), it first took to the air on 9 February 1980. Incorporating an AN/APG-71 radar, NVG-compatible cockpits, a twin IRST/TV undernose pod, enhanced AAM capability and AN/ALR-67 RWR equipment, the F-14D is a far more capable aircraft, but only 37 of the planned 127 were built before the programme was cancelled. Current US plans are to retrofit some 200 F-14A/Bs with some F-14D features.

Specification (F-14D)

Powerplant: two 120.1 kN (27,000 lb st) General Electric F110-GE-400 afterburning turbofans

Dimensions: length 19.10 m (62 ft 8 in); height 4.88 m (16 ft 0in); wing span (fully swept) 10.15 m (33 ft 3½ in), (fully spread) 19.54 m (64 ft 1½ in)

Weights: empty 18,951 kg (41,780 lb); MTOW 33,724 kg (74,349 lb)

Performance: max level speed at high altitude ('clean') 1,997 km/h (1,241 mph); max rate of climb at sea level +9,145 m (30,000 ft)/min; service ceiling +16,150 m (53,000 ft)

Armament: one General Electric M61A-1 20 mm Vulcan cannon with 675rds, four AIM-54C Phoenix long-range AAMs, four AIM-7 Sparrow medum-range AAMs, four AIM-9 Sidewinder short-range AAMs, Rockeye and CBU-59 cluster bombs, two auxiliary fuel tanks

IAI Kfir

Israel developed the Kfir for ground attack missions

The Israeli Kfir (Lion Cub) combines the Dassault Mirage III's airframe with the powerful J79 afterburning turbojet engine acquired when the USA supplied Israel with F-4 Phantom IIs. This airframe/engine combination first flew on 19 October 1970, followed the next year by a J79-powered Nesher (an Israeli copy of the Mirage 5). These aircraft formed the basis of a production aircraft first flown during 1972. Featuring enlarged engine air intakes, a dorsal fin ram air inlet, a longer nose, revised cockpit and indigenous avionics, the baseline Kfir was produced in small numbers, most of the 27 built subsequently being upgraded to C1 standard, the most obvious changes being the addition of canard foreplanes above each air intake and a small strake either side of the nose. These greatly enhanced combat manoeuvrability and low-speed handling. Ironically, 25 Kfir C1s were later leased to the US Navy and USMC as F-21As.

Further upgrading in the mid-1970s led to the Kfir C2, new features including a dogtooth wing leading-edge, new radar, a multi-mode navigation and weapons delivery system. IAI built a total of 185 Kfir C2s and TC2 two-seat trainers, small numbers of which have been exported to Colombia and Ecuador. The 1980s saw extant C2/TC2s further upgraded and 125 Kfirs remain in service today.

Specification (Kfir-C7)

Powerplant: one 83.40 kN (18,750 lb st) licence-built General Electric J79-J1E afterburning turbojet

Dimensions: length 15.65 m (51 ft 4¼ in); height 4.55 m (14 ft 11¼ in); wing span 8.22 m (26 ft 11½ in)

Weights: empty, equipped 7,285 kg (16,060 lb); MTOW 16,500 kg (36,376 lb)

Performance: max level speed at sea level 1,389 km/h (863 mph), at 10,975 m (36,000 ft) 2,440 km/h (1,516 mph); max rate of climb at sea level 14,000 m (45,930 ft)/min; service ceiling 17,680 m (58,000 ft)

Armament: two DEFA 553 30 mm cannon with 140 rpg; up to 6,085 kg (13,415 lb) of ordnance including Shafrir 2 and Python 3 AAMs, GBU-13 LGBs, AGM-65 Maverick ASMs, Durandal anti-runway bombs, Mk82/84 free-fall bombs, cluster bombs, napalm tanks, ECM pods, auxiliary fuel tanks

Kamov Ka-50 Werewolf

The Ka-50 is the Russian army's latest gunship

The Ka-50 is the world's first single-seat close-support
helicopter. The first flight took place on 27 July 1982,
but it was 1989 before a photograph of the new design
was available for publication in the West. By then a
competitor - the Mil-28 - was flying, but competitive
evaluation of the two designs led to the Ka-50
emerging as victor. Protected by some 350kg (159lb) of
armour to counter enemy fire up to 20mm, the Ka-50
retains Kamov's trademark coaxial contra-rotating
three-blade main rotor. Heat suppressors cut the
exhaust plume from each of the widely-spaced TV3-
117VK turboshafts, and the pilot sits in a double-wall
steel cockpit behind bulletproof flat-screen glazing.
Should the pilot have to make an emergency exit, the

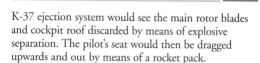

K-37 ejection system would see the main rotor blades
and cockpit roof discarded by means of explosive
separation. The pilot's seat would then be dragged
upwards and out by means of a rocket pack.

Unlike its Western counterparts, the Ka-50 relies on
other aircraft to locate and designate targets, the logic
being that this reduces the helicopter's vulnerability to
attack by minimizing its exposure.

The single-barrel 30mm gun is mounted on the
starboard side of the fuselage and can be traversed
through +5-6deg in azimuth and -30deg in elevation.
Pods at each wing-tip house ECM equipment and
chaff/flare dispensers.

Specification

Powerplant: two 1,633 kW (2,190 shp) Klimov TV3-117VK
turboshafts

Dimensions: length (fuselage) 13.50 m (44 ft 3½ in), (rotors
turning) 16.0 m (52 ft 6 in); height 4.93 m (16 ft 2 in); main rotor
diameter 14.5 m (47 ft 7 in)

Weights: normal take-off weight 9,800 kg (21,605 lb); MTOW
10,800 kg (23,810 lb)

Performance: max speed (in level flight) 310 km/h (193 mph), (in
shallow dive) 350 km/h (217 mph); vertical rate of climb at 2,500
m (8,200 ft) 600 m (1,970 ft)/min;

Armament: one 2A42 30 mm cannon, four 20-round B-8 80 mm
unguided rocket pods, 12 laser-guided ASMs, 23 mm gun pods,
AAMs, weapons dispensers

Lockheed P-3 Orion

US P3s tracked Russian subs throughout the Cold War

When the US Navy issued a requirement for a land-based anti-submarine and maritime patrol warplane to replace the Lockheed P2V Neptune, Lockheed though that the ideal type could be evolved from its Model 88 Electra four-turboprop civil transport. The US Navy concurred, and the resulting YP3V prototype made its first flight on 25 November 1959 as a derivative of the Electra with a shorter fuselage accommodating a weapon bay for 2,722kg (6,000lb) of the type's maximum warload of 9,072kg (20,000lb). The P-3A entered service in August 1962 with a powerplant of four 3,356ekW (4,500ehp)

T56-A-10W turboprops, and production amounted to 157 aircraft There followed the P-3B (125 aircraft). The final new-build model for the US Navy is the current P-3C with the same powerplant as the P-3B, but with new sensors. Many of the P-3C fleet of 267 aircraft have been cycled through five update programmes. Some of the aircraft have been converted to transport, staff transport, oceanographic survey, weather reconnaissance and Elint aircraft. Other aircraft have been exported to Australia, Canada, Iran, the Netherlands, New Zealand, Norway, Pakistan, Portugal, South Korea and Spain, and the type is built under licence in Japan by Kawasaki.

Specification (P-3C Orion)

Powerplant: four 3,66 kW (4,910 ehp) Allison T56-A-14 turboprops

Dimensions: length 35.61m (116 ft 10 in); height 10.27 m (33 ft 8½ in); wing span 30.37 m (99 ft 8in)

Weights: take-off ('clean') 61,235 kg (135,000 lb); MTOW 64,410 kg (142,000 lb)

Performance: max level speed at 4570m (15,000 ft) 761 km/h (457 mph); service ceiling 8625 m (28,300 ft)

Armament: 9072 kg (20,000 lb) of disposable stores, including nuclear or conventional depth charges, mines, bombs, torpedoes, rockets, AGM-84 Harpoon AShMs

Lockheed F-104 Starfighter

The F-104 is finally retiring from frontline service

Experience in the Korean War convinced Lockheed that the US Air Force would soon need an exceptionally fast-climbing interceptor, and the company therefore designed the F-104 as what was immediately dubbed a 'manned missile' for its massive fuselage and diminutive straight wings; the airframe was completed by a T-tail. The XF-104 prototype first flew in March 1954, and the F-104A initial model entered service in 1958. Production totalled 153 aircraft, some of which were later exported after the USAF's need for the type diminished. There followed 77 examples of the F-104C fighter. Tandem two-seat trainer versions of these two USAF models were built as the F-104B and F-104D respectively. West German requirement for a multi-role warplane were met by the F-104G. Production for West Germany and other European countries totalled 1,127 aircraft, and further aircraft were built under licence in Canada (200 CF-104s) and Japan (210 F-104Js). Variants of this basic theme were the RF-104G reconnaissance model and TF-104G tandem two-seat trainer that was also built in the USA, Canada and Japan as the F-104D, CF-104D and F-104DJ respectively. These aircraft are disappearing from final service in the mid-1990s, leaving the somewhat revised Alenia F-104S (245 aircraft for Italy and Turkey) as the last operational model.

Specification (F-104G Starfighter)

Powerplant: one 70.28 kN (15,800 lb st) General Electric J79-GE-11A turbojet

Dimensions: length 16.69 m (54 ft 9i n); height 4.11m (13 ft 6in); wing span (excluding tip stores) 6.68 m (21 ft 11 in)

Weights: take-off ('clean') 9,838 kg (21,690 lb); MTOW 13,054 kg (28,779 lb)

Performance: max level speed at 10,975 m (36,000 ft) Mach 2.2 or 2,333 km/h (1,450 mph); service ceiling 17,680 m (58,000 ft)

Armament: one 20 mm M61A1 Vulcan six-barrel cannon with 750 rounds; 1,995 kg (4,310 lb) of disposable stores, including nuclear weapons, ASMs, AAMs, and free-fall bombs, on external hardpoints

Lockheed AC-130A/E/H/U Spectre

AC-130s continue to support US Special Operations

Early US experience in the Vietnam War highlighted the need for an aerial platform to provide concentrated fire power against ground targets. This led to the development of the aerial gunship, initial offerings being armed versions of the C-47 and C-119. As the need for such aircraft increased, the C-130A Hercules was selected for modification as an aerial gunship.

Four 20mm cannon, four 7.62mm Miniguns, FLIR target-acquisition, a searchlight, improved sensors and direct-view image intensifiers characterized the AC-130A Spectre, 14 of which were built and used operationally over Vietnam from late 1968 onwards. Their success led to the conversion of 11 C-130Es with better armour protection, more ammunition and enhanced avionics. From 1973, these aircraft were upgraded to AC-130H standard with more powerful engines, a 105mm howitzer, two 40mm and two 20mm cannon, and the addition of IR/LLLTV sensors, laser target designator, sideways-looking HUD for aiming at night while orbiting the target and an IFR capability.

Development of an all-new version, the AC-130U, began in mid-1987, the first post-conversion flight taking place on 20 December 1990. While the 105mm howitzer and 40mm cannon remain, the two 20mm have been replaced by a pair of 25mm six-barrel Gatling guns. Two targets can be fired upon simultaneously, and all operations are controlled and monitored by operators at seven computer consoles in the cabin-mounted battle-management centre.

Specification (AC-130U)

Powerplant: four 3,362 kW (4,508 shp) Allison T56-A-15 turboprops

Dimensions: length 29.79 m (97 ft 9 in); height 11.66 m (38 ft 3 in); wing span 40.41m (132 ft 7 in)

Weights: empty, operational 34,536 kg (75,743 lb); MTOW 79,380 kg (175,000 lb)

Performance: max cruising speed at optimum altitude 556 km/h (345 mph); max rate of climb at sea level 579 m (1,900 ft)/min; service ceiling 10,060 m (33,000 ft)

Armament: one 105 mm howitzer, one Bofors 40 mm gun, one General Electric GAU-12/U 25 mm six-barrel gun with 3,000rds

Lockheed/Boeing F-22A Rapier

The F-22 will maintain America's lead in fighter design

In the early 1980s the USAF announced a programme to develop a replacement for the F-15C Eagles now in service. In 1986 the USAF selected two designs for prototype testing: the Lockheed/Boeing YF-22A and the Northrop YF-23A. The YF-22A and its Pratt & Whitney F119 powerplant won in April 1991, leading to a contract for 11 (now nine) flying prototypes (seven single-seat F-22As and two two-seat F-22Bs), a static airframe and a fatigue test airframe.

Although the two YF-22As were constructed largely of metal, some 28% of the F-22A/B airframe will comprise composites. The angular design incorporates low observables in its construction and configuration, the latter having been refined somewhat for production aircraft. Manoeuvrability is enhanced by a triplex digital FBW flight control system and two-dimensional thrust-vectoring exhaust nozzles, while the advanced technology reheat powerplants allow the aircraft to operate in Supercruise (supersonic cruise without afterburner augmentation).

Such performance combined with the in-built stealth characteristics will give the F-22A an important 'first look-first kill' capability. Three internal bays (one in the side of each engine intake duct and one in the ventral bay) can carry AAMs or AAMs/PGMs respectively. The USAF wanted some 648 F-22s, but is now due to get 442. First flight is scheduled for early 1997, with deliveries of production F-22As to run from 2000-2011.

Specification (F-22, provisional)

Powerplant: two 155 kN (35,000 lb st) Pratt & Whitney F119-PW-100 afterburning turbofans

Dimensions: length 18.92 m (62 ft 1 in); height 5.00 m (16 ft 5 in); wing span 13.56 m (44 ft 6 in)

Weights (estimated): empty 13,608 kg (30,000 lb); MTOW approx 27,716 kg (60,000 lb)

Performance: max level speed (in supercruise) Mach 1.58, (with afterburning, at 9,150 m (30,000 ft)) Mach 1.7, (at sea level) 1,482 km/h (921 mph); service ceiling 15,240 m (50,000 ft)

Armament: one M61A2 20 mm gun, two JDAM 1000 PGMs, four AIM-9 and four AIM-120 AMRAAM AAMs

Lockheed F-117A Nighthawk

The F-117 shot to fame during the 1991 Gulf War

Popularly referred to as the 'Stealth Fighter', the F-117A Nighthawk is a precision attack aircraft designed to be 'invisible' to radar. Constructed primarily of aluminium, the F-117A's fuselage comprises flat panels known as facets mounted on the aircraft's subframe, their purpose being to reflect radar energy away from the transmitter itself, thus denying the operators a viable 'return'. All surfaces are coated with various radar absorbent materials. All doors and panels have serrated edges to further minimize radar reflection. Grid covers on the intakes and the use of narrow-slot 'platypus' exhausts surrounded by heat-absorbing tiles further reduce the chances of IR detection.

Ahead of the flat-plate five-piece cockpit glazing is a FLIR sensor, recessed in a mesh-covered housing; in the forward starboard underfuselage there is a retractable DLIR and laser designator. These sensors are used in conjunction with LGBs, two of which can be carried in the double-section weapons bay. Alternatively, AAMs can be carried. That the weapons load is relatively modest is testimony to the faith put in the low-observable technology - a faith borne out by the fact that 42 F-117As flew just 2% of the combat sorties mounted during the Gulf War, yet accounted for some 40% of the strategic targets attacked in Iraq.

Specification

Powerplant: two 48.0 kN (10,800 lb st) General Electric F404-GE-F1D2 non-afterburning turbofans

Dimensions: length 20.08 m (65 ft 11 in); height 3.78 m (12 ft 5 in); wing span 13.20 m (43 ft 4 in)

Weights: empty (estimated) 13,381 kg (29,500 lb); MTOW 23,814 kg (52,500 lb)

Performance: max level speed 1,040 km/h (646 mph); mission radius (unrefuelled, 2,268 kg (5,000 lb) weapon load) 1,056 km (656 miles)

Armament: up to 2,268 kg (5,000 lb) of GBU-10 Paveway II/GBU-27 Paveway III LGBs, AGM-65 Maverick ASMs, AGM-88 HARMs, AIM-9 AAMs, gun pods

Lockheed S-3 Viking

The S-3 is a carrier-based anti-submarine aircraft

As the need for a replacement for its faithful turboprop-powered Grumman S-2 Tracker ASW aircraft became more apparent, the US Navy invested its hopes in Lockheed's high-wing, twin-turbofan S-3A design, the first flight of which (by one of eight pre-production YS-3As) took place on 21 January 1972. Successful testing led to a contract for 179 production S-3As, now named Viking, the first of which entered operational service in July 1974.

The Viking is equipped with state-of-the-art data processing and detection systems, the core of which is the AN/ASQ-81 tail-mounted extendible MAD 'stinger'. In addition there is an AN/APS-116 high-resolution maritime search radar, an OR-89 FLIR, and 60 sonobuoys located in the aft fuselage. The S-3B model, developed in the early 1980s, has enhanced ESM, a new AN/ARR-78(V) sonobuoy receiver system and it can carry two AGM-84 Harpoon missiles. Some 160 S-3As were upgraded to S-3B standard. Those S-3As not upgraded were the small number of US-3A COD aircraft, these having had their ASW systems stripped out to create space within the fuselage for passengers and mail.

The most recent model to enter service is the ES-3A Elint platform, first flown in 1991, distinguishable by the numerous external antennae and a prominent dorsal 'shoulder pack'.

Specification (S-3A)

Powerplant: two 41.26 kN (9,275 lb st) General Electric TF34-GE-2 non-afterburning turbofans

Dimensions: length 16.26 m (53 ft 4 in); height 6.93 m (22 ft 9 in); wing span 20.93 m (68 ft 8 in)

Weights: empty 12,088 kg (26,650 lb); MTOW 23,832 kg (52,540 lb)

Performance: max level speed at sea level 814 km/h (506 mph); patrol speed at optimum altitude 296 km/h (184 mph); max rate of climb at sea level 1,280 m (4,200 ft)/min; service ceiling +10,670 m (35,000 ft)

Armament: Mk82 free-fall bombs, Mk54 or Mk57 depth bombs, Mk53 mines, Mk46 torpedoes, Mk36 destructors, cluster bombs, rocket pods, flare launchers, two auxiliary fuel tanks

McDonnell Douglas A-4 Skyhawk

Skyhawks provided vital ground support in Vietnam

First flown on 22 June 1954, the 'Bantamweight
Bomber' or 'Scooter' was designed to meet US
Navy/Marine Corps carrierborne light attack
requirements of the time. Early models were powered
by a Wright J65 turbojet, different models of which
powered the A-4A/B/Cs. In the mid-1960s came the
J52-powered A-4E with five stores pylons, followed
by the A-4F which introduced the distinctive dorsal
avionics 'hump'. A-4Fs for export include the A-4G,
A-4H, A-4K and A-4KU. Seventy-seven A-4Cs were
rebuilt for the US Naval Reserve as A-4Ls, these also
featuring the dorsal 'hump', while a further 66 A-4Cs
were rebuilt in the late 1960s as the A-4P and A-4Q
for the Argentine Air Force and Navy respectively.

146

The final export customer for rebuilt second-hand A-4Cs was Singapore; these A-4Cs had two 30mm guns in place of the US Navy's 20mm weapons.

Utilization of the 20% more powerful J52-P-408A turbojet effectively led to a second generation of A-4s. First in the new family was the 'humped' A-4M. Built at the same time was the A-4N for the IDF/AF, featuring uprated avionics, decoy flare dispensers and a cockpit HUD. South-East Asia is home to many surviving A-4s, those in Singapore including A-4S-1s and SUs. Malaysia's A-4PTMs (ex-A-4C/Ls) can fire AGM-65 ASMs and AIM-9 AAMs, while New Zealand's has added new avionics and radar to its A-4Ks and TA-4K trainers.

Specification (A-4S-1)

Powerplant: one 48.04 kN (10,800 lb st) General Electric F404-GE-100D non-afterburning turbofan

Dimensions: length 12.72 m (41ft 8½ in); height 4.57m (14 ft 11¾ in); wing span 8.38 m (27 ft 6 in)

Weights: empty, equipped 4,649 kg (10,250 lb); MTOW 10,206 kg (22,500 lb)

Performance: max level speed at sea level 1,128 km/h (701mph); max rate of climb at sea level 3,326 m (10,913 ft)/min; service ceiling 12,190 m (40,000 ft)

Armament: two Mk 12 20 mm cannon with 200rpg; up to 3,720 kg (8,200 lb) of ordnance including ASMs, bombs, AIM-9 Sidewinder AAMs, auxiliary fuel tanks

McDonnell Douglas AH-64A Apache

Built to attack Soviet tanks, the AH-64 is being upgraded

First flown on 30 September 1975, the Apache is a
tandem two-seater heavily armoured to withstand
strikes from 23mm ammunition. The crew (CP/G
front, pilot rear) sit in cockpits fitted with armoured
shields resistant to 12.7mm strikes. The rugged
airframe and energy-absorbing landing gear offer a
95% chance of impact survival at rates of descent up to
12.8m (42ft) per second. The 'Black Hole' IR
suppression system protects the Apache from heat-
seeking missiles .As to the Apache's offensive
capabilities, the key to success lies in the nose-mounted
TADS/PNVS (Target Acquisition and Designation

Sight/Pilot Night Vision Sensor). The PNVS comprises a wide-angle FLIR sensor that provides the pilot with a real-time thermal image of the land ahead. Thus the pilot can fly NOE mission profiles at night or in adverse weather. Once in the target area, the TADS (FLIR, day TV camera, laser spot tracker and laser rangefinder/designator) allows the CP/G to locate, designate and track targets by day or night via direct-view optics. Target range is verified by the active laser rangefinder/designator, a laser spot allowing the Hellfire missile(s) to home accurately.

Specification

Powerplant: two 1,265 kW (1,696 shp) General Electric T700-GE-701 turboshafts

Dimensions: length 17.76 m (58 ft 3 in); height (over tail rotor) 4.30 m (14 ft 1¼ in); wing span (over weapon racks) 5.82 m (19 ft 1 in); main rotor diameter 14.63 m (48 ft 0 in)

Weights: empty 5,165 kg (11,387 lb); primary mission gross weight 6,552 kg (14,445 lb); MTOW 9,525 kg (21,000 lb)

Performance (at 6,552 kg (14,445 lb)): never-exceed speed 365 km/h (227 mph); max level speed 293 km/h (182 mph); max vertical rate of climb at sea level 762 m (2,500ft)/min; service ceiling 6,400 m (21,000 ft)

Armament: one M230 30 mm Chain Gun with up to 1,200rds; up to 16 Hellfire ATGMs, 76 70 mm (2.75 in) FFARs in four rocket pods, two AIM-9 Sidewinder or four Sidearm/Stinger/Mistral AAMs, four auxiliary fuel tanks

McDonnell Douglas AH-64D

The AH-64D has a very distinctive mast-mounted sight

Confusingly, the AH-64D Longbow Apache designation also applies to a model of the Apache attack helicopter noticeably different to the upgraded AH-64A. The most obvious difference is the Westinghouse Longbow mast-mounted radar which, combined with Hellfire ATGMs equipped with radio frequency seeker heads, forms an integrated fire control radar and missile system capable of locating, tracking and despatching targets in the air and on the ground through levels of smoke, rain and fog that would prove too much for IR sensors. The Longbow radar atop the mast scans through 360deg for aerial targets or 270deg for ground targets, presenting up to 256 targets on the gunner's Tactical Situation Display. Once a target is selected, the Hellfire offers a 'fire-and-forget' capability as it can lock on to its target before launch, or can launch and subsequently lock on to the target's coordinates once in flight.

Deliveries of production-standard Longbow Apaches are scheduled to start during 1997, with current US Army plans calling for some 227 AH-64As. Export potential is currently focused on the UK and the Netherlands, though no firm orders have been placed by either nation at the time of writing.

Specification

Powerplant: two 1,447 kW (1,940 shp) General Electric T700-GE-701C turboshaft engines

Dimensions: length 17.76 m (58 ft 3 in); height 4.95 m (16 ft 3 in); wing span (over empty weapon racks) 5.82 m (19 ft 1 in) main rotor diameter 14.63 m (48 ft 0 in);

Weights: empty 5,352 kg (11,800 lb); primary mission gross weight 6,552 kg (14,445 lb); MTOW 10,107 kg (22,823 lb)

Performance: max level speed 261 km/h (162 mph); max rate of climb 474 m (3,090 ft)/min; service ceiling 3,800 m (12,480 ft)

Armament: one M230 30 mm Chain Gun with up to 1,200rds; up to 16 Hellfire ATGMs on four wing pylons, 76 70 mm (2.75 in) FFARs in four rocket pods, two AIM-9 Sidewinder or four Sidearm/Stinger/Mistral AAMs, four auxiliary fuel tanks

McDonnell Douglas F-4 Phantom II

F-4s served with distinction in many western air forces

One of the finest warplanes ever designed, and built in larger numbers than any other Western warplane since World War II, the angular but aggressive-looking Phantom II was initially conceived as a carrierborne attack warplane, placed in initial production as a carrierborne fleet defence fighter, and then transformed into a multi-role fighter for land-based as well as carrierborne use. The carrierborne models for the US Navy lacked any inbuilt gun. The land-based models for the US Air Force included the F-4C minimum-change version of the F-4B, the F-4D with APQ-109 radar, the F-4E with an inbuilt cannon, APQ-120 radar and a number of aerodynamic enhancements, and the F-4G 'Wild Weasel' defence suppression conversion of the F-4E with specialized avionics and provision for anti-radar weapons.

US Marine Corps and USAF reconnaissance models were the RF-4B and RF-4C respectively, the latter being exported as the RF-4E in a variant based on the F-4E. Other export models were the F-4F for West Germany and the F-4K and F-4M carrierborne and land-based models for the UK.

The F-4E was also built under licence in Japan by Mitsubishi; and Germany and Israel have both undertaken major upgrades of their F-4Fs and F-4Es for the air superiority and ground-attack roles respectively.

Specification (F-4E Phantom II)

Powerplant: two 79.62 kN (17,900 lb st) General Electric J79-GE-17A turbojets

Dimensions: length 19.20 m (63 ft 0 n); height 5.02 m (16 ft 5½ in); wing span 11.77 m (38 ft 7½ in)

Weights: take-off ('clean') 18,818 kg (41,487 lb); MTOW 28,030 kg (61,795 lb)

Performance: max level speed at 10,975 m (36,000 ft) Mach 2.17 or 2301 km/h (1,430 mph); service ceiling 17,905 m (58,750 ft)

Armament: one 20 mm M61A1 Vulcan six-barrel cannon with 640 rounds; 7,257 kg (16,000 lb) of disposable stores, including nuclear weapons, ASMs, AAMs, free-fall or guided bombs, cluster bombs and ECM pods, carried on nine external hardpoints.

McDonnell Douglas F-15C Eagle

The F-15C is still the fighter by which others are judged

Designed as replacement for the F-4 Phantom II in the land-based air superiority role and first flown in YF-15 prototype form on 27 July 1972, the F-15 Eagle is another type that has matured as a superb multi-role fighter. The type was planned with a large wing and a thrust/weight ratio exceeding unity for unrivalled climb rate, and from the beginning has been notable for the advanced nature of its avionics and the excellent fields of vision provided for the pilot. The first two models to enter service were the F-15A single-seater and TF-15A (later F-15B) combat-capable tandem two-seater: both had APG-63 radar and Pratt & Whitney F100-P-100 afterburning turbofans, and production totalled 366 and 58 respectively for the US Air Force plus 19 and two respectively for the Israeli Air Force.

Delivered from June 1979, the definitive variants are the F-15C single-seat and F-15D two-seat models with APG-70 radar (with a programmable digital signal processor, synthetic-aperture ground mapping and track-while-scan air-to-air capability), an uprated powerplant, and provision for low-drag conformal packs carrying fuel and fitted with tangential attachments for weapons. Both types remain in production for the USAF, Israel and Saudi Arabia, and are built under licence in Japan by Mitsubishi.

Specification (F-15C Eagle)

Powerplant: two 105.73 kN (23,770 lb st) Pratt & Whitney F100-P-220 turbofans

Dimensions: length 19.43 m (63 ft 9 in); height 5.63 m (18 ft 5½ in); wing span 13.05 m (42 ft 9½ in)

Weights: take-off ('clean') 20,244 kg (44,630 lb); MTOW 30,845 kg (68,000 lb)

Performance: max level speed at 10,975 m (36,000 ft) more than Mach 2.5 or 2,655 km/h (1,665 mph); service ceiling 18,290 m (60,000 ft)

Armament: one 20 mm M61A1 Vulcan six-barrel cannon with 940 rounds; 10,705 kg (23,600 lb) of disposable stores, including nuclear weapons, ASMs, AAMs, free-fall or guided bombs, cluster bombs, dispenser weapons, rocket launchers, napalm tanks, drop tanks and ECM pods, carried on nine external hardpoints.

155

McDonnell Douglas F-15E Strike Eagle

The F-15E Strike Eagle is an outstanding fighter-bomber

Although similar in general appearance to the F-15B/D Eagle two-seat trainers, a total of 18 weapons hardpoints identify the F-15E Strike Eagle as an altogether more ferocious bird. Its existence can be traced back to the early 1980s, when the USAF formally identified a need for an aircraft to replace ageing F-4Es and supplement its F-111E/F force.

Successful trials with a converted TF-15A led to the go-ahead in early 1984 for full-scale development of this potent dual-role fighter. The maiden flight of the first production F-15E took place on 11 December 1986, and the first of just over 200 operational aircraft for the USAF were delivered in 1988/89. Success during the Gulf War of 1991 led to export orders from

Israel (21 F-15Is) and Saudi Arabia (72 downgraded F-15Ss), these to be delivered during the second half of the 1990s.

The F-15E's primary mission is air-to-ground strike, and to this end a wide variety of guided/unguided weapons can be carried on underwing and centreline pylons, and tangential stores carriers fitted to the CFTs. Up to eight AAMs can be carried for use in the air superiority role, backed up by a six-barrel cannon in the starboard wing-root for close-range work.

Specification

Powerplant: two 129.45 kN (29,100 lb st) Pratt & Whitney F100-PW-229 afterburning turbofans
Dimensions: length 19.43 m (63 ft 9 in); height 5.63 m (18 ft 6 in); wing span 13.05 m (42 ft 10 in)
Weights: operating empty 14,515 kg (32,000 lb); MTOW 36,741 kg (81,000 lb)
Performance: max level speed at high altitude ('clean') +2,655 km/h (1,650 mph); max rate of climb at sea level +15,240 m (50,000 ft)/min; max range 4,455 km (2,762 miles)
Armament: one 20 mm M61A1 gun with 512rds; up to 11,000 kg (24,250 lb) of ordnance including B57/-61 nuclear bombs, GBU-10/-12/-15/-24 LGBs, Rockeye and CBU-52/-58/-71/-87/-89/-90/-92/-93 CBUs, Mk82/84 'iron' bombs, AGM-65 Maverick ASMs, AGM-88 HARMs, AIM-7 Sparrow/AIM-120 AMRAAM medium-range AAMs, AIM-9 Sidewinder short-range AAMs, three auxiliary fuel tanks

McDonnell Douglas F/A-18 Hornet

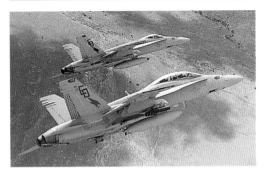

The F/A-18 Hornet doubles as fighter and strike aircraft

When its YF-17 prototype lost to the General Dynamics YF-16 in the US Air Force's light-weight fighter competition, Northrop decided to collaborate as junior partner with McDonnell Douglas in the development of a navalized derivative to meet a US Navy requirement for a carrierborne warplane to replace the McDonnell Douglas F-4 Phantom II and Vought A-7 Corsair II in the fighter and attack roles. The resulting prototype made its first flight on 18 November 1978, and although initial thought had been given to separate F-18 and A-18 fighter and attack variants, it was soon realized that one type could undertake both roles, merely by changes in the computer software. Considerable development had to be undertaken, and it was November 1983 before the Hornet entered service in its initial F/A-18A single-seat and F/A-18B combat-capable tandem two-seat variants. Deliveries of these two model to the US Navy and US Marine Corps totalled 371 and 39 respectively, and other aircraft were sold to Australia, Canada and Spain.

The current production models are the F/A-18C single- and FA/-18D two-seat models, which are being bought by the USN and USMC and also by Finland, Kuwait and Switzerland.

Specification (F/A-18C Hornet)

Powerplant: two 71.17 kN (16,000 lb st) General Electric F404-GE-400 turbofans

Dimensions: length 17.07 m (56 ft 0 in); height 4.66 m (15 ft 3½in); wing span (excluding tip stores) 11.43 m (37 ft 6 in)

Weights: take-off ('clean') 16,652 kg (36,710 lb); MTOW 22,328 kg (49,224 lb)

Performance: max level speed at high altitude more than Mach 1.8 or 1,913 km/h (1,189 mph); service ceiling about 15,240 m (50,000 ft)

Armament: one 20 mm M61A1 Vulcan six-barrel cannon with 570 rounds; 7,711 kg (17,000 lb) of disposable stores, including AAMs, ASMs, anti-ship missiles, free-fall and guided bombs, cluster bombs, dispenser weapons, napalm tanks and ECM pods, carried on nine external hardpoints.

McDonnell Douglas MD500/530 Defender

For its size, the MD500 carries an incredible armament

Winner of the US Army's 1960 competition to find a new light observation helicopter, Hughes' YHO-6 prototype flew for the first time on 27 February 1963. The production-standard OH-6A Cayuse entered service in 1965, and soon the distinctive tadpole-like shape and brittle rasp of its rotors were well-known. Many are still operational, although mostly with second-line units. Current plans are to upgrade them to OH-6B standard which includes a more powerful engine and a chin-mounted FLIR.

Hughes (later McDonnell Douglas Helicopters) went on to develop the world's most successful family of light military helicopters. The Hughes 500, devised for the civilian market, spawned a host of military-configured offspring including anti-tank versions armed with TOW missiles and naval variants equipped for ASW with underslung torpedoes and a MAD (Magnetic Anomaly Detector). US Army Special Forces fly specialised models including EH-6Es (sigint and command roles) and MH-6Es to land troops behind enemy lines. A second generation is based on the streched MD500E civilian helicopter, military variants being known collectively as the MD500MG Defender.

Specification (500MG Defender)

Powerplant: one 313 kW (419 hp) Allison 250-C20B turboshaft

Dimensions: length 7.29 m (23 ft 11 in); height to top of rotor head 2.62 m (8ft 7 in); height with MMS 3.41 m (11ft 2 in); width over skids 1.96 m (6ft 5 in)

Weights: empty, equipped 898 kg (1,979 lb); MTOW 1,406 kg (3,100 lb)

Performance: max speed at sea level 241 km/h (150 mph); max rate of climb 626 m (2,055 ft)/min; max endurance, no reserves 1 h 56 min

Armament: TOW 2, 7.62 mm or 12.7 mm machine gun pods, 2.75 in (69 mm) rockets in 7- or 12-tube launchers, Stinger AAMs

Mikoyan-Gurevich MiG-21 'Fishbed'

The MiG-21 was the mainstay of Communist air forces

The MiG-21 'Fishbed' was schemed as a clear-weather interceptor and first flew in 1957 in the form of the Ye-6 prototype. The type entered production in 1958 with the Tumanskii R-11 turbojet, but reached its definitive form in the following year with the MiG-21F 'Fishbed-C' with an engine uprated to 56.39 kN (12,676lb st). This paved the way for the MiG-21PF 'Fishbed-D' of 1960 with the uprated R-11F engine and an enlarged inlet to allow the incorporation of R1L 'Spin Scan-A' radar in the centrebody, the MiG-21PFS 'Fishbed-E' with blown flaps and a ventral pod carrying a 23mm GSh-23L two-barrel cannon, and the MiG-21PFM 'Fishbed-F' incorporating all the sequential improvements of earlier models. The second generation MiG-21PFMA 'Fishbed-J' featured additional fuel, 'Jay Bird' radar, an internal GSh-23L cannon, and four rather than two underwing hardpoints for air-to-surface, as well as air-to-air weapons, which made this a dual-role fighter. The first of the third-generation types was the MiG-21bis 'Fishbed-L' with a stronger yet lighter airframe and updated avionics. The final MiG-21Mbis 'Fishbed-N' has the R-25 turbojet and provision for more modern weapons. There are also three tandem two-seat trainers models, and in the mid-1990s there is a market for the revision of in-service aircraft with modern western avionics.

Specification (MiG-21bis 'Fishbed-N')

Powerplant: one 69.61kN (15,650 lb st) Tumanskii R-25-300 turbojet

Dimensions: length 12.285 m (40 ft 3½ in); height 4.10 m (13 ft 5½ in); wing span 7.154 m (23 ft 5½ in)

Weights: take-off ('clean') 8,725 kg (19,235 lb); MTOW 10,400 kg (22,925 lb)

Performance: max level speed at 13,000 m (42,650 ft) Mach 2.05 or 2175 km/h (1,353mph); ceiling 17500 m (57,400 ft)

Armament: one 23 mm GSh-23L two-barrel cannon with 200 rounds; 1,500 kg (3,307 lb) of disposable stores, including AAMs, single large-calibre rockets, bombs, drop tanks and ECM pods, carried on four external hardpoints

Mikoyan-Gurevich MiG-23 'Flogger'

MiG-23s still serve with the former Warsaw Pact nations

Schemed in the early 1960s as successor to the MiG-21, the MiG-23 'Flogger' was developed with a variable-geometry wing to provide better payload/range and field performance. The first prototype flew on 10 June 1967, and MiG-23M 'Flogger-B' production aircraft entered service in 1973 with the Khachaturov R-29-300 turbojet and the S-23D-Sh 'High Lark' radar. Another Khachaturov turbojet, the R-35, is used in late-production models such as the MiG-23ML 'Flogger-G' lightened version of the MiG-23M, the MiG-23P 'Flogger-G' with a digital autopilot slaved to a ground-controller system, and the MiG-23MLD 'Flogger-K' updated version of the MiG-23ML with dogtoothed wing leading edges and a number of avionics upgrades. The third type of engine used is Tumanskii R-27F2M-300 turbojet, which powers the MiG-23UB 'Flogger-C' combat-capable tandem two-seat trainer and the MiG-23MF and MiG-23MS 'Flogger-E' export versions of the MiG-23M. with downgraded avionics.

The fourth engine type is the 112.76kN (25,350lb st) Lyul'ka AL-21F-300 turbojet, which powers the MiG-23B 'Flogger-F' attack model which paved the way for the MiG-23B series. The MiG-23 was built in large numbers up to the late 1980s, and is still in very widespread service.

Specification (MiG-23ML 'Flogger-G')

Powerplant: one 127.49 kN (28,660 lb st) Khachaturov R-35-300 turbojet

Dimensions: length 16.71m (54 ft 10 in); height 4.82 m (15 ft 9½ in); wing span 13.965 m (45 ft 10 in) spread and 7,779 m (25 ft 6½ in) swept

Weights: take-off ('clean') 14,700 kg (32,405 lb); MTOW 17,800 kg (39,250 lb)

Performance: max level speed at high altitude Mach 2.35 or 2,500 km/h (1,553 mph); service ceiling 18,500 m (60,700 ft)

Armament: one 23 mm GSh-23L two-barrel cannon with 200rounds; 3,000 kg (6,614 lb) of disposable stores, including nuclear weapons, AAMs, ASMs, free-fall and guided bombs, cluster bombs, dispenser weapons, and ECM pods

Mikoyan-Gurevich MiG-25

MiG-25s were built to shoot down US strategic bombers

Realizing the threat posed by the American Mach 3-capable XB-70 Valkyrie strategic bomber, the Soviet Union urgently set about development of a new interceptor. The Valkyrie never progressed beyond the prototype stage, but the Soviet interceptor project went ahead regardless. Actually, a Ye-155R-1 recce prototype made the first flight on 6 March 1964, followed by the Ye-155P-1 interceptor prototype on 9 September 1964. The new design from the Mikoyan-Gurevich stable had high-set wings, twin outward-canted tailfins, and very large intakes feeding air to two Mikulin R-15B-300 turbojets. The Ye-155R-1 led in turn to the MiG-25R recce model, soon redesignated MiG-25RB after a bombing capability was added in 1970. Meanwhile the Ye-155P-1 led to the first fighter production model, the MiG-25P armed with up to four underwing AAMs.

The MiG-25P led to the MiG-25PD, the definitive interceptor model. Powered by two uprated R-15BD-300s and equipped with a Sapfir-25 radar. As for the recce-bombers, the MiG-25RBK was a derivative of the MiG-25RB minus cameras and with different Elint equipment. The MiG-25RB also gave rise to the MiG-25BM, tailored for defence-suppression tasks. Armed with four Kh-58 stand-off anti-radiation missiles for use against SAM radar sites, it has a lengthened nose with a dielectric panel on either side and ECM in place of the recce equipment module.

Specification (MiG-25PD)

Powerplant: two 110 kN (24,700 lb st) Soyuz/Tumansky R-15BD-300 afterburning turbojets

Dimensions: length 23.82 m (78 ft 1¾ in); height 6.10 m (20 ft ¼ in); wing span 14.02 m (45 ft 11¾ in)

Weights: 'clean', max internal fuel 34,920 kg (76,985l b); with four R-40 AAMs and max internal fuel 36,720 kg (80,950 lb)

Performance: max level speed at 13,000 m (42,650 ft) 3,000 km/h (1,865 mph), at sea level 1,200 km/h (745 mph); time to 20,000 m (65,600 ft) 8.9min; service ceiling 20,700 m (67,900 ft); endurance 2h 5min

Armament: R-23 (two), R-40 (two) R-60 (four) and R-73A (four) AAMs

Mikoyan-Gurevich MiG-27

The MiG-27 strike aircraft is a much improved MiG-23B

Experience with attack-configured MiG-23s led to the development of a dedicated ground attack derivative in the 1970s to support Warsaw Pact ground forces. The result was the MiG-27, slightly longer and significantly heavier than the MiG-23 and with a nav/attack system tailored to the ground attack role.

Production-standard MiG-27s started to enter service during the second half of the 1970s, the first model being very similar to the MiG-23BM but with fixed engine air intakes and simpler, two-position (on/off) afterburner nozzles. Further improvements led in turn to the MiG-27K, based on the MiG-23BK. The PrNK-23K nav/attack system enables automatic flight control, weapons release and gun firing, as well as allowing the aircraft to navigate automatically to the target at night and/or in cloud and carry out an attack using preprogrammed coordinates.

The MiG-27D is noted for improvements to its aerodynamic form and a revised nose incorporates a noticeable upper 'lip' extending over an enlarged window for a new Klen laser rangefinder. Several sub-variants of the MiG-27D exist, each representing a progressive update of the basic model. Production of new-build MiG-27s ended in the then Soviet Union in the mid-1980s.

Specification (MiG-27)

Powerplant: one 112.7 kN (25,335 lb st) Soyuz/Khachaturov R-29B-300 afterburning turbojet

Dimensions: length 17.08 m (56 ft ¼ in); height 5.00 m (16 ft 5 in); wing span 13.97m (45 ft 10 in) fully spread, 7.78 m (25 ft 6¼ in) fully swept

Weights: empty 11,908 kg (26,252 lb); MTOW 20,300 kg (44,750 lb)

Performance: max level speed at 8,000 m (26,250 ft) 1,885 km/h (1,170 mph); max rate of climb at sea level 12,000 m (39,370 ft)/min; service ceiling 14,000 m (45,900 ft)

Armament: one GSh-6-30 30 mm twin-barrel cannon with 260rds; up to 4,000 kg (8,818 lb) of 50-500 kg (110-1,102 lb) bombs, tactical nuclear bombs, Kh-23/-29 ASMs, 57 mm/240 mm rocket packs, napalm containers, R-3S/-13M AAMs, three auxiliary fuel tanks

Mikoyan-Gurevich MiG-29 'Fulcrum'

Malaysia is the most recent customer for the MiG-29

Resulting from a 1972 requirement for a fighter to succeed the Mikoyan-Gurevich MiG-21 and MiG-23 and also the Sukhoi Su-15 and Su-17 in the counter-air and attack roles, the MiG-29 resembles the McDonnell Douglas F-15 in basic configuration, but has approximately the size of the smaller McDonnell Douglas F/A-18. A fly-by-wire control system was considered but not adopte. However, despite this fact, the MiG-29 had revealed extraordinary agility, among the several facets that make this one of the most important fighters anywhere in the world. The first example flew on 6 October 1977, and the type entered service in 1984 in its initial MiG-29 'Fulcrum-A' form that was delivered in three forms (with ventral fins, with the ventral fins replaced by fin fillets containing flare launchers, and with extended-chord rudders) as the last technical problems were overcome.

The MiG-29UB 'Fulcrum-B' combat-capable tandem two-seat operational conversion trainer version of the 'Fulcrum-A' was produced in limited numbers in parallel with the 'Fulcrum-A'; and the 'Fulcrum-C' single-seater introduced a deepened dorsal fairing, which allowed much of the upgraded avionics to be removed from the lower fuselage to allow an increase in internal fuel tankage. The MiG-29K 'Fulcrum-D' was developed as a carrierborne version for use on the USSR's planned aircraft-carrier force, but was rejected in favour of a Sukhoi Su-27.

Specification (MiG-29 'Fulcrum-A')

Powerplant: two 81.40 kN (18,300 lb st) Sarkisov RD-33 turbofans

Dimensions: length 17.32m (56 ft 10 in); height 4.73 m (15 ft 6½in); wing span 11.36 m (37 ft 3½ in) **Weights:** take-off ('clean') 15,240 kg (33,600 lb); MTOW 18,500 kg (40,785 lb)

Performance: max level speed at high altitude Mach 2.3 or 2,445 km/h (1,520 mph); service ceiling 17,000 m (55,775 ft)

Armament: one 30 mm GSh-30-1 cannon with 150 rounds; 3,000 kg (6,614 lb) of disposable stores including AAMs, ASMs, free-fall or guided bombs, cluster bombs, dispenser weapons, rocket launchers, drop tanks and ECM pods, carried on six external hardpoints.

Mikoyan-Gurevich MiG-31

The MiG-31 is a heavily-armed long range interceptor

Whilst the MiG-25 was designed to meet the threat posed by the high-flying XB-70 Valkyrie strategic bomber, the MiG-31 long-range supersonic interceptor evolved to counter the threat posed by enemy aircraft and cruise missiles flying high-speed, low-level mission profiles. The design was first flown in prototype (Ye-155MP) form on 16 September 1975, the aircraft concerned being a converted MiG-25MP.

Although generally similar in appearance to the MiG-25, the MiG-31 differs in several key respects, resulting in what is virtually an all-new aircraft. The airframe, for example, has been strengthened to absorb the rigours of low-level supersonic flight, power for which comes from two D-30F6 turbofans which in turn have necessitated the use of larger air intakes and extended exhaust nozzles. New main landing gear units have also been adopted, these featuring tandem offset wheels which retract into bays in the intake trunks.

Up front, the nose radome houses a Zaslon phased-array fire control radar capable of tracking 10 targets at ranges up to 120km (75 miles), and attacking four of them simultaneously. Successful tracking and engagement is the responsibility of the WSO, seated behind the pilot in a cockpit with limited glazing. Weaponry at his disposal includes four radar-homing long-range AAMs located beneath the fuselage and up to six AAMs.

Specification (MiG-31)

Powerplant: two 151.9 kN (34,170 lb st) Aviadvigatel D-30F6 afterburning turbofans

Dimensions: length 22.69 m (74 ft 5¼ in); height 6.15 m (20 ft 2¼ in); wing span 13.46 m (44 ft 2 in)

Weights: empty 21,825 kg (48,115 lb); MTOW 46,200 kg (101,850 lb)

Performance: max level speed at 17,500 m (57,400 ft) 3,000 km/h (1,865 mph), at sea level 1,500 km/h (932 mph); time to 10,000 m (32,800 ft) 7min 54sec; ceiling 20,600 m (67,600 ft)

Armament: one GSh-6-23 23 mm six-barrel gun with 260rds, four R-33 AAMs, two R-40T AAMs, four R-60 AAMs

Mil Mi-8 'Hip'

Mi-8s served as transports and gunships in the Afghan war

Basically a turbine-engined derivative of the Mi-4 'Hound' for civil as well as military applications, the Mi-8 first flew in 1961 in its 'Hip-A' prototype form with a four-blade main rotor powered by a 2,013kW (2,700shp) Soloviev turboshaft, but the type was then revised at the 'Hip-B' prototype with a five-blade main rotor and a powerplant of two Isotov TV2-117 turboshafts. The first production model was the 'Hip-C' that introduced Doppler navigation and provision for heavy armament used in the defence suppression role. The Mi-8 has proved itself an excellent assault and utility transport helicopter and, as a result, the type is still built in its upgraded Mi-17 form with an uprated powerplant of two 1,454kW (1,950shp) TV3-117MT turboshafts.

The Mi-8 has been produced in a number of forms, of which the totals remain unknown. Among the military variants of the Mi-8 and Mi-17 are the Mi-8 'Hip-D' for the airborne communications role, Mi-8 'Hip-E' derivative of the 'Hip-C' with still heavier air-to-surface armament in the form of 192 rockets in six packs and four AT-2 'Swatter' anti-tank missiles, Mi-8 'Hip-F' export version of the 'Hip-E' with AT-3 'Sagger' anti-tank missiles, Mi-9 'Hip-J' airborne communications model, Mi-17 'Hip-H' upgraded version of the 'Hip-C' and the Mi-8 'Hip-J' and 'K' ECM models.

Specification (Mi-8 'Hip-C')

Powerplant: two 1,270 kW (1,703 shp) Isotov TV2-117A turboshafts

Dimensions: length 25.24 m (82 ft 9½ in); height 5.65 m (18 ft 6½ in); main rotor diameter 21.29 m (69 ft 10½ in)

Weights: take-off ('clean') 11,100 kg (24,470lb); MTOW 12,000 kg (26,455 lb)

Performance: max level speed at 1,000 m (3,280 ft) 260 km/h (161mph); service ceiling 4,500 m (14,760 ft)

Armament: this is carried on two outriggers extending from the fuselage sides, and comprises four launchers each carrying 32.55 mm (2.17 in) rockets.

Mil Mi-24 'Hind'

The Mi-24 is heavily armoured against ground fire

Developed in the mid-1960s as a high-performance helicopter for the military role and still in production, the Mi-24 has been operated in two distinct forms as an assault helicopter and, in a much adapted form, as a gunship helicopter. The type was originally conceived as an assault helicopter with retractable tricycle landing gear, a flightcrew of three or four and accommodation for eight soldiers, and stub wings allowing rockets for the suppression of ground defences in the landing zone to be carried. The first model, probably a pre-production model, was the 'Hind-B' with straight, rather than anhedralled, wings carrying four hardpoints. The following 'Hind-A' was the first production variant and, in addition to an anhedralled wing with six hardpoints, relocated the tail rotor to the starboard side of the fin. The 'Hind-C' was built in small numbers without the nose gun and wing tip missile-launch rails for use as a dual-control trainer.

The Hind-D' has much improved armament, including a four-barrel machine gun in a chin turret, as well as a comprehensive array of offensive and defensive sensors, the latter being used to trigger decoy flares. The 'Hind-E' has a 30mm twin-barrel fixed cannon on the starboard side of the forward fuselage, AT-6 'Spiral' anti-tank missiles, an IR jammer and engine shrouds. The 'Hind-G' is a specialized radiation sampling variant.

Specification (Mi-24 'Hind-D')

Powerplant: two 1640 kW (2,200 shp) Isotov TV3-117 turboshafts

Dimensions: length, rotor turning 21.50 m (70 ft 6½ in); height 6.50 m (21 ft 4 in); main rotor diameter 17.00 m (55 ft 9 in)

Weights: take-off ('clean') 11,000 kg (24,250 lb); MTOW 12,500 kg (27,557 lb)

Performance: max level speed at optimum altitude 335 km/h (208 mph); service ceiling 4,500 m (14,750 ft)

Armament: one 12.7 mm (0.5 in) four-barrel machine gun in the chin turret; 1,500 kg (3,307 lb) of disposable weapons, including anti-tank missiles, mine dispensers, rocket launchers, grenade-launcher pods, cannon pods and machine gun pods, carried on six external hardpoints.

Mil Mi-28 'Havoc'

The Mi-28 is built to fight both tanks and helicopters

First flown in November 1982 for service from the later 1990s, the two-seat Mi-28 'Havoc-A' is clearly derived from earlier Mil helicopters (including the dynamic system of the Mi-24 driving a new five-blade articulated main rotor), but has adopted the US practice of a much smaller and slimmer fuselage for increased manoeuvrability and reduced battlefield vulnerability. The Mi-28 thus bears a passing resemblance to the Hughes AH-64A Apache in US Army service. Among its operational features are IR suppression of the podded engines' exhausts, IR decoys, upgraded steel/titanium armour, optronic sighting and targeting systems for use in conjunction with the undernose 30mm cannon and disposable weapons (including AAMs) carried on the stub wing hardpoints, and millimetric-wavelength radar.

The type clearly possesses an air-combat capability against other battlefield helicopters, and other notable features are a far higher level of survivability and the provision of a small compartment on the left-hand side of the fuselage, probably for the rescue of downed aircrew.. In 1993 it was announced that the type is in fact to be procured for Russian service alongside the Ka-50. Due to fly in 1995 for a possible service debut in 1997, the Mi-28N 'Havoc-B' is a night and adverse-weather version.

Specification (Mi-28 'Havoc')

Powerplant: two 1,640 kW (2,200 shp) Isotov TV3-117 turboshafts

Dimensions: length, fuselage excluding tail rotor 16.85 m (55 ft 3½in); height 4.81m (15 ft 9½ in); main rotor diameter 17.20 m (56ft 5in)

Weights: empty 7,000 kg (15,432lb); MTOW 10,400 kg (22,928 lb)

Performance: max level speed at sea level 300 km/h (186 mph); service ceiling not revealed

Armament: one 30 mm 2A42 cannon with 300 rounds; 16 modified AT-6 'Spiral' anti-tank missiles in four four-tube boxes, or four UV-20-57 launchers each carrying 20 55 mm (2.17 in) rockets, or four UV-20-80 launchers each carrying 20 80 mm (3.15 in) rockets, or four AAMs in two twin launchers or a combination of these weapon types, carried on four external hardpoints.

Mitsubishi F-1

Mitsubishi F1s are usually fitted with anti-ship missiles

A Japanese requirement for a supersonic trainer to help pilots' transition from the F-104J to the F-4EJ led to development of the two-seat Mitsubishi F-2. First flown in prototype (XT-2) form on 20 July 1971, the design owed much to the Anglo-French SEPECAT Jaguar, and its success led in turn to plans for a single-seat attack derivative with a limited counter-air capability. Two T-2s were duly converted and redesignated FST-2, the first flight of the new fighter prototype taking place on 3 June 1975.

Using the same airframe, engines and systems as the T-2, the first of 77 production F-1s entered JASDF service in April 1978. The most obvious difference is the deletion of the rear cockpit, the area being used to house avionics instead. The F-1's principal mission is anti-shipping, and the J/AWG-12 radar provides both air-to-air and air-to-surface operating modes and is compatible with the F-1's primary weapon, the ASM-1 AShM. The ASM-1 has a range of some 50km (31 miles) and uses active radar guidance to home onto enemy warships.

While the ASM-1 is the F-1's principal armament, the F-1's five hardpoints can also carry a variety of bombs, rockets and AAMs up to 2,722kg (6,000lb). A cannon is located in the lower port forward fuselage along with 750 rounds of 20mm ammunition.

Specification
Powerplant: two 32.49 kN (7,305 lb st) Ishikawajima-Harima TF40-IHI-801 (licence-built Rolls-Royce/Turbomeca Adour Mk 801A) afterburning turbofans
Dimensions: length 17.86 m (58 ft 7 in); height: 4.39 m (14 ft 5 in); wing span 7.88 m (25 ft 10¼ in)
Weights: empty, equipped 6,358 kg (14,017 lb); MTOW 13,700 kg (30,203 lb)
Performance: max level speed ('clean' at 10,975 m (36,000 ft)) 1,700 km/h (1,056 mph); max rate of climb at sea level 10,670 m (35,000 ft); service ceiling 15,240 m (50,000 ft)
Armament: one JM61 Vulcan 20 mm multi-barrel cannon with 750rds; up to 2,721 kg (6,000 lb) of ordnance including ASM-1 AShMs, 227 or 340 kg (500 or 750 lb) bombs, JLAU-3A 70 mm rockets, RL-7 70 mm rockets, RL-4 125 mm rockets, AIM-9L AAMs, three auxiliary fuel tanks

Mitsubishi FS-X

The advanced FSX is about to begin development trials

Delays caused by workshare disputes between Japan and the USA, and escalating costs, have taken their toll on the development of the FS-X, the JASDF's next single-seat fighter-bomber; but the first of four prototypes (two single-seat FS-Xs and two twin-seat TFS-Xs) was finally rolled-out by Mitsubishi on 12 January 1995. Developed as a replacement for the 77 Mitsubishi F-1s still in service, the FS-X is based on the Lockheed F-16C Block 50 Fighting Falcon - a fact plain to see in illustrations of the new aircraft. The first flight is scheduled for September 1995, the four prototypes being joined in the test programme by two static airframes for fatigue testing. Deliveries of pre-production aircraft are scheduled to begin in early 1996, with entry into JASDF service following shortly thereafter.

Although Mitsubishi is in overall control of the project, various Japanese and American subcontractors have been assigned aspects of the design and development work. Differences from the F-16 include a new cold-cured composite wing of Japanese design that is greater in both span and area, a longer radome and forward fuselage to accommodate the active phased-array radar and other avionics, and a longer mid-fuselage and shorter tailpipe for the F110-GE-129 IPE afterburning turbofan powerplant. In the cockpit, state-of-the-art displays include a Yokogawa LCD multi-function display and a Shimadzu holographic display.

JASDF trials are expected to run from 1996-99 and current plans call for at least 72 aircraft.

Specification

Powerplant: one 129.0 kN (29,000 lb st) General Electric F110-GE-129 afterburning turbofan

Dimensions: length 15.27m (50 ft 1¼ in); height 4.97 m (16 ft 3¾ in); wing span (over AAM rails) 11.13 m (36 ft 6¼ in)

Weights: empty 9,525 kg (21,000 lb); MTOW (with external stores) 22,100 kg (48,722 lb)

Armament: one M61A1 Vulcan 20 mm multi-barrel cannon; five external stores points for Mitsubishi XASM-2 ASMs, AAM-3 and AIM-7 Sparrow AAMs

NAMC Q-5/A-5

The A-5 has been sold to Pakistan and North Korea

Designed to meet a 1950s requirement for a supersonic attack aircraft, the Qiangjiji-5 (Attack aircraft 5) first flew on 4 June 1965. Development of the Q-5 led to the Q-5I. A major change involved deletion of the bomb bay in favour of more fuel, thus extending the aircraft's range. Strengthened landing gear and an extra pair of underfuselage hardpoints were fitted, and some Navy aircraft were fitted with Doppler radar compatible with their role as sea-skimming delivery platforms for the C-801 AShM and underfuselage torpedoes. Further enhancement of the Q-5I's offensive capabilities, in the form of an extra pair of underwing hardpoints and a new gun/bomb-sighting system, led in the mid-1980s to the Q-5IA; the addition of RWR resulted in the Q-5II.

The Q-5IA accounted for the type's first export success, namely 40 for North Korea. Keen to win further orders, Nanchang offered the Q-5IA for further export as the A-5C (Q-5III). Incorporating 32 modifications of the Q-5 plus upgraded avionics and an added AIM-9 AAM capability, the A-5C was ordered by Bangladesh (20) and Pakistan (52).

Development of the Q-5II in cooperation with Alenia of Italy subsequently produced the all-weather A-5M. Myanmar has ordered 24 such aircraft.

Specification (A-5C)

Powerplant: two 31.87 kN (7,165lb st) Shenyang WP6 afterburning turbojets

Dimensions: length 16.25 m (53 ft 4 in); height 4.52 m (14 ft 10 in); wing span 9.70 m (31 ft 10 in)

Weights: empty 6,494 kg (14,317 lb); MTOW 12,000 kg (26,455 lb)

Performance: max level speed at 11,000 m (36,000 ft) 1,190 km/h (740 mph); max rate of climb at 5,000 m (16,400 ft) 4,980-6,180 m (16,340-20,275 ft)/min; service ceiling 15,850 m (52,000 ft)

Armament: two 23 mm cannon each with 100rds; up to 2,000 kg (4,410 lb) of bombs, C-801 AShMs, rockets, PL-2/-2B/-7/AIM-9/R.530 Magic AAMs, auxiliary fuel tanks

Northrop F-5E Tiger II

Many US allies continue to fly the 20-year old Tiger II

Evolved as an upgraded version of the F-5A Freedom fighter light tactical fighter offering better performance and weapons capability through the introduction of an uprated powerplant and a number of aerodynamic enhancements, the F-5E Tiger II first flew in March 1969 and soon revealed that the design team had fully met its brief, as well as providing greater fuel capacity, improved field performance and a more modern integrated fire-control system into which the Emerson APQ-153 or APQ-159(V) lightweight multi-mode radars could be introduced. Final development proceeded without undue problem, and the F-5E entered service in April 1973 at the start of a major production programme that saw the delivery of more than 1,400 aircraft to many of the USA's allies before the close of the programme in 1986.

The F-5F Tiger II has its fuselage lengthened by 1.08 m (3ft 6½in). The type has Emerson APQ-157 radar and only one 20mm cannon, but can be fitted with a system for control of the AGM-65 Maverick air-to-surface missile. The RF-5E TigerEye is the reconnaissance model with its fuselage lengthened to make space for a pallet able to accommodate several combinations of sensors. The F-5E series is still in large-scale service, and there is a considerable market for the evolution of upgraded versions.

Specification (F-5E Tiger II)

Powerplant: two 22.24 kN (5,000 lb st) General Electric J85-GE-21 turbojets

Dimensions: length 14.45 m (47 ft 4½ in); height 4.06 m (13 ft 4 in); wing span 8.13 m (28 ft 8 in)

Weights: empty 4,410 kg (9,723l b); MTOW 11,214 kg (24,722 lb)

Performance: max level speed at 10,975 m (36,000 ft) Mach 1.64 or 1,743 km/h (1,083 mph); service ceiling 15,790 m (51,800 ft)

Armament: two 20 mm M39A2 cannon with 280 rounds per gun; 3,175 kg (7,000 lb) of disposable stores, including AAMs, ASMs, free-fall bombs, drop tanks and ECM pods, carried on seven external hardpoints.

Northrop Grumman B-2A Spirit

The 'flying wing' design of the B-2 hides it from radar

Revealed on 22 November 1988, the B-2A Spirit low-observable (stealth) strategic bomber is the result of a programme begun in 1978. The first of six prototypes made its maiden flight on 17 July 1989, and a test programme is scheduled for completion during 1997. One operational USAF unit, the 509th BW, operates the Spirit, the first example for operational use having been delivered in1993.

The B-2A is a blended flying wing with straight leading-edges and a 'sawtooth' trailing-edge. A centrebody smoothly contoured into the upper wing surfaces contains the two-man crew compartment and two weapons bays, while the all-important contouring extends to the engine bays. The exhausts for the engines are deliberately positioned well forward of the wing trailing-edge, this to help reduce the heat signature.

The 15 Block 10 production B-2As have no RWR or TFR, but these features will be added from 1997 onwards. Block 20 aircraft will incorporate a GPS-aided targeting system. Block 30 aircraft will have fully automated mission planning and a sat/nav communications system. The planned total of just 22 B-2As is a far cry from the USAF's original plans to acquire 133 such aircraft.

Specification

Powerplant: four 84.5 kN (19,000 lb st) General Electric F118-GE-110 non-afterburning turbofans

Dimensions: length 21.03 m (69 ft 0 in); height 5.18 m (17 ft 0 in); wing span 52.43 m (172 ft 0 in)

Weights: empty 45,360-49,900 kg (100,000-110,000 lb); MTOW 170,550 kg (376,000 lb)

Performance: max level speed at high level 764 km/h (475 mph); service ceiling 15,240 m (50,000 ft); range at MTOW with 16,919 kg (37,300 lb) warload 11,667 km (7,250 miles), at 162,386 kg (358,000 lb) take-off weight with 10,886 kg (24,000 lb) warload 12,223 km (7,595 miles); range with one aerial refuelling 18,520 km (11,508 miles)

Armament: 22,680 kg (50,000 lb) including 16 B61/B83 free-fall nuclear bombs, 80 Mk82 454 kg (1,000 lb) or 16 Mk84 908kg (2,000 lb) bombs, 36 M117 340.5 kg (750 lb) fire bombs, 36 cluster bombs, and 80 Mk36 454 kg (1,000 lb) sea mines

Panavia Tornado ADV

The ADV was designed for the air defence of the UK

Foreseeing the need to replace its Lightnings and
Phantoms in the air defence and air superiority roles,
the RAF decided to produce a derivative of the
Tornado IDS dedicated to air defence of the UK,
protection of NATO's northern and western
approaches, and the long-range protection of UK
naval forces at sea. The result was the Tornado ADV,
full-scale development of which was authorized on 4
March 1976. The first of three prototypes took to the
air on 27 October 1979, with deliveries of 18
production Tornado F.2s completed by October 1985.
The following month, the F.3 prototype took to the

air on its maiden flight.

Although it retains a high commonalty of parts with the Tornado IDS, the ADV has a longer, more slender nose, housing an AI Mk 24 Foxhunter multi-mode pulse-Doppler radar capable of detecting targets at up to 185km (115 miles) and tracking several simultaneously. The fuselage has also been extended aft of the rear cockpit to create enough length to enable carriage of two tandem pairs of semi-recessed Sky Flash AAMs beneath the fuselage. In addition, up to four short-range AAMs can be carried on two inboard underwing pylons. New-generation AAMs can also be carried, but the forward-mounted starboard 27mm cannon has been deleted.

Specification

Powerplant: two 73.5 kN (16,520 lb st) Turbo-Union RB199-34R Mk 104 afterburning turbofans

Dimensions: length 18.68 m (61ft 3½ in); height 5.95 m (19 ft 6¼ in); wing span (fully swept) 8.60 m (28 ft 2½ in), (fully spread) 13.91 m (45 ft 7½ in)

Weights: operational empty 14,500 kg (31,790 lb); MTOW 27,986 kg (61,700 lb)

Performance: max level speed at 10,975 m (36,000 ft) 2,338 km/h (1,453 mph); operational ceiling 21,335 m (70,000 ft)

Armament: one IWKA-Mauser 27 mm cannon, four Sky Flash medium-range AAMs, two AIM-9L Sidewinder short-range AAMs, two auxiliary fuel tanks

Panavia Tornado IDS

GR Mk1s will soon be upgraded with better electronics

Still one of the most important warplanes in the world despite the age of its basic design, the Tornado was schemed in the late 1960s as the two-seat Multi-Role Combat Aircraft able to deliver a substantial warloads with pinpoint accuracy over long ranges after take-off from a short runway. The type was developed in a collaborative programme by British, Italian and West German interests for production by the industries of these three countries. The basic design was based on the use of two advanced turbofan engines offering a high power/weight ratio and excellent fuel economy, a compact airframe with a variable-geometry wing fitted with extensive high-lift devices and controlled by a fly-by-wire system, and an advanced nav/attack system incorporating advanced

radar (with search, ground-mapping and terrain-following capabilities), inertial navigation and HUD. The first Tornado flew on 14 August 1974, and the type entered service in July 1980.

The baseline version is the Tornado IDS that is known to the RAF as the Tornado GR.Mk 1 and also serves, with slightly different avionics, in the German, Italian and Saudi Arabian air forces as well as the German Navy. The RAF also operates a few aircraft adapted to the Tornado GR.Mk 1A standard with a multi-sensor reconnaissance suite.

Specification (Tornado GR.Mk 1)
Powerplant: two 74.73 kN (16,800 lb st) Turbo-Union RB.199-34R Mk 103 turbofans
Dimensions: length 16.72 m (54 ft 10½ in); height 5.95 m (28 ft 2½ in); wing span 13.91 m (45 ft 7½ in) spread and 8.60 m (28 ft 2½ in) swept
Weights: take-off ('clean') 20,410 kg (44,996 lb); MTOW about 27,215 kg (60,000 lb)
Performance: max level speed at 11,000 m (36,090 ft) more than Mach 2.2 or 2,337 km/h (1,453 mph); service ceiling more than 15,240 m (50,000 ft)
Armament: two 27 mm Mauser BK27 cannon with 180 rounds per gun; 9,000 kg (19,840 lb) of disposable stores including nuclear weapons, AAMs, ASMs, anti-radar missiles, anti-ship missiles, free-fall and guided bombs, cluster bombs, dispenser weapons, drop tanks and ECM pods, carried on seven external hardpoints.

193

Rockwell OV-10 Bronco

The US Marines have operated the Bronco for 30 years

A much underestimated warplane designed for the battlefield reconnaissance and counter-insurgency roles, the Bronco first flew during July 1965. A production order was then placed for the OV-10A variant and the first of 271 production aircraft (114 for the US Marine Corps and 157 for the USAF) flew in August 1967, and the Bronco performed admirably in the battlefield reconnaissance, light attack, FAC and helicopter escort roles. By the early 1990s, the USAF's force had been reduced to some 20 aircraft, some of the aircraft having been exported (to Morocco and the Philippines) and others earmarked for conversion to OV-10D standard as supplements to the US Marine Corps' force of this important variant.

The OV-10B was produced for West Germany as a target-towing version of the OV-10A with no armament or stub wings. Other export versions of the OV-10A include the OV-10C, E and F (40 for Thailand, and 16 each for Venezuela and Indonesia). The most advanced variant is the OV-10D conversion of the OV-10A for the USMC's' Night Observation Surveillance role with more powerful engines. A 20 mm General Electric M197 three-barrel cannon and 1,500 rounds can be carried on the centreline hardpoint when the sponsons are omitted.

Specification (OV-10A Bronco)

Powerplant: two 533 kW (715 ehp) Garrett T76-G-416/417 turboprops

Dimensions: length 12.67 m (41ft 7in); height 4.62 m (15 ft 2in); wing span 12.19 m (40 ft 0 in)

Weights: take-off ('clean') 4,494 kg (9,908 lb); MTOW 6,552 kg (14,444 lb)

Performance: max level speed at sea level 452 km/h (281mph); service ceiling 7,315 m (24,000 ft)

Armament: four 7.62 mm (0.3 in) M60C machine guns with 500 rounds per gun; 1,633 kg (3,600 lb) of disposable stores, including AAMs, free-fall bombs, cluster bombs, dispenser weapons, rocket launchers, cannon pods, machine guns and drop tanks, carried on seven external hardpoints.

Rockwell B-1B Lancer

Transition to low level has proved difficult for the B1

The B-1 was designed as a supersonic strategic bomber able to penetrate Soviet defences and launch stand-off nuclear weapons. Accordingly, the B-1A (of which four prototypes were built, the first taking to the air on 23 December 1974) featured a long, slim fuselage blended with a variable-geometry wing with a maximum sweep of 67deg.

The four F101 turbofan engines were paired in underwing pods, with plans calling for each engine to have a variable inlet. This feature was subsequently abandoned when the B-1's mission changed to a low-level high-subsonic profile, a task allocated to the production B-1B of which 100 were built. First flown on 18 October 1984, the B-1B began to enter service with the USAF's Strategic Air Command in July 1985.

Changes from the original B-1A include strengthened landing gear, redesigned wing gloves, fixed engine inlets and significant use of composite materials in the aircraft's construction to reduce weight. RAM is also used to cover much of the airframe, thus reducing the big bomber's radar cross-section to less than one-hundredth that of a B-52 Stratofortress.

Specification

Powerplant: four 136.92 kN (30,780 lb st) General Electric F101-GE-102 afterburning turbofans

Dimensions: length 44.81 m (147 ft 0 in); height 10.36 m (34 ft 10 in); wing span (fully swept) 23.84 m (78 ft 2½ in), (fully spread) 41.67 m (136 ft 8½ in)

Weights: empty, equipped 87,091 kg (192,000 lb); MTOW 216,365 kg (477,000 lb)

Performance: max level speed ('clean' at high altitude) 1,324 km/h (823 mph); penetration speed at 61m (200 ft) 965 km/h (600 mph)

Armament: up to 34,020 kg (75,000 lb) of ordnance including B-61/-83 free-fall nuclear bombs, Mk82 227 kg (500 lb) conventional free-fall bombs, AGM-69A SRAM-As, AGM-86B ALCMs, AGM-86C ALCMs, Mk36 227 kg (500 lb) mines

197

Saab J35 Draken

The Draken's double delta profile is unmistakable

Some 40 years after the prototype first took to the air, the Saab Draken still has a sleek, futuristic appearance about it. The initial production model (J35A) entered service in March 1960 powered by a licence-built Rolls-Royce Avon fitted with a more efficient afterburner. The J35A was superseded by the longer J35B, this model also introducing the distinctive twin tailwheels to aid aerodynamic braking on landing. Both models served as air-defence interceptors. The more powerful RM6C engine, revised inlets, more fuel and upgraded radar characterized the J35D, which in turn lent itself to further development resulting in the J35E recce variant and the J35F, the final air-defence variant. This model had one of the two 30mm cannon deleted, the emphasis now being on the use of licence-built Falcon IR-homing AAMs. Relatively few Drakens now remain in Swedish service; those that do are single-seat J35Js (upgraded J35Fs with two extra underwing hardpoints and better radar, IR sensor, avionics and cockpit displays) and two-seat Sk35Cs. The latter aircraft has also found its way abroad as part of export orders, primarily to other Scandinavian air forces, the result of Saab's decision to offer the Saab 35X (J35F with enhanced attack capabilities) to overseas customers including Denmark, Finland and Austria.

Specification (J35J)

Powerplant: one 78.51 kN (17,650 lb st) Volvo Flygmotor RM6C afterburning turbojet

Dimensions: length 15.35 m (50 ft 4 in); height 3.89 m (12 ft 9 in); wing span 9.40 m (30 ft 10 in)

Weights: empty, equipped 8,250 kg (18,188 lb); MTOW 12,270 kg (27,050 lb)

Performance: max level speed ('clean' at 10,975 m (36,000 ft)) +2,126 km/h (1,321 mph); max rate of climb at sea level 10,500 m (34,450 ft)/min

Armament: one Aden M/55 30 mm cannon; up to 2,900 kg (6,393 lb) of ordnance including Rb24, Rb27 and Rb28 AAMs, up to four auxiliary fuel tanks

Saab AJ 37 Viggen

The Viggen doubles as ground attack/naval strike aircraft

Developed for the Swedish Air Force in the 1960s, the Viggen was the first canard-equipped fast jet to enter service. This feature counteracts the worst problems associated with the delta wing configuration when operating low and slow. It enhances the aircraft's STOL characteristics - important given the need to be able to operate off-base from relatively short stretches of motorway in times of crisis. Such operations are helped by the high take-off thrust and reverse thrust on landing offered by the Viggen's RM8 afterburning turbofan, although this is also quite a thirsty powerplant. Naturally, provision has been made for the carriage of external auxiliary fuel tanks.

The first version of this multi-role aircraft to be

developed was the AJ 37 attack fighter, which entered service on 21 June 1971. Capable of operating against surface targets on land or sea, the AJ 37's primary armament is the RB04E AshM, RB05A ASM/AShM and Rb75 licence-built Maverick TV-guided ASM. Seven pylons are available for ordnance and/or external fuel tanks weighing up to 6,000kg (13,200lb), other weapons including rocket launchers and gun pods, and RB24/RB28 AAMs for self-defence.

Front-line variants of the AJ 37 are the SF 37 (all-weather recce by day or night) and SH 37 (all-weather sea surveillance with a secondary maritime strike role).

Specification

Powerplant: one 11,790 kg (25,970 lb) Volvo Flygmotor RM8A afterburning turbofan
Dimensions: length 16.30 m (53 ft 5¾ in); height 5.60 m (18 ft 4½ in); wing span 10.60 m (34 ft 9¼ in); canard span 5.45 m (17 ft 10½ in)
Weights: empty 12,250 kg (27,000 lb); normal take-off 17,000 kg (37,479 lb)
Performance: max speed at high altitude 2,124 km/h (1,320 mph); rate of climb at sea level 12,200 m (40,000 ft)/min; service ceiling 18,290 m (60,000 ft)
Armament: up to 6,000 kg (13,228 lb) of ordnance including Rb04, Rb05 and Rb75 ASMs, Rb24 and Rb28 AAMs, six-round 135 mm (3.5 in) rocket pods, 30 mm Aden gun pods

Saab JA 37 Viggen

The JA 37 usually carries 6 anti-aircraft missiles

The last member of the Viggen family to enter service with the Royal Swedish Air Force, the JA 37, is a dedicated interceptor with a secondary ground attack capability, designed and procured as a replacement for J 35 Drakens. Externally, the JA 37 differs little from the AJ 37 and its single-seat derivatives, other than a taller fin (adapted from the Sk 37 two-seat trainer) with a distinctive kinked tip, an extra elevon actuator beneath each wing (taking the total to four per wing) and a ventral 30mm gun pack; but internally it is quite a different beast.

First flown on 27 September 1974, the JA 37 has

better stability at all operational heights and attitudes. A boost in thrust to the tune of 8% is the result of a new high-pressure turbine and four-nozzle combustion burner system. The JA 37 also features a lookdown/shootdown radar, a central digital computer, radar display and RWR, a HUD for the pilot and digital automatic flight control.

The first JA 37s entered service with the Royal Swedish Air Force in 1978, with an eventual total of 149 such interceptors being procured, making it the most numerous model in the Viggen family. The recent introduction of the multi-role JAS 39 Gripen marks the beginning of the end for the JA 37.

Specification

Powerplant: one 125.04 kN (28,100 lb st) Volvo Flygmotor RM8B afterburning turbofan

Dimensions: length 16.40 m (53 ft 9¾ in); height 5.90 m (19 ft 4¼ in); wing span 10.60 m (34 ft 9¼ in); canard span 5.45 m (1 7ft 10½ in)

Weights: normal take-off (intercept) 15,000 kg (33,069 lb), (attack) 20,500 kg (45,194 lb)

Performance: max level speed ('clean' at 10,975 m (36,000 ft)) +2,126 km/h (1,321mph); max speed at sea level 1,470 km/h (910 mph)

Armament: one 30 mm cannon with 150rds; up to 5,897 kg (13,000 lb) of ordnance including up to six Rb71 and Rb74 AAMs, four six-round 135 mm (5.3 in) rocket pods

Saab JAS 39 Gripen

Saab maintains its track record with the agile Gripen

Proud of its track record in fighter design, Sweden again chose to look to Saab to find a new single-seat multi-purpose fighter to replace all four operational models of the Saab Viggen and the final few Drakens. The result was the JAS 39 Gripen, a small, lightweight design which first flew on 8 December 1988.

'JAS' - Jagt Attack Spaning (Fighter-Attack-Reconnaissance) - gives a clue as to the Gripen's three-fold responsibilities in the modern-day Royal Swedish Air Force. The aircraft itself is of close-coupled delta canard configuration, and light but extremely strong composites account for some 30% of the airframe structure. Configuring the aircraft for different mission profiles is facilitated by easily programmable software and associated systems, information being presented to the pilot on three head-down CRT multi-function displays and a wide-angle HUD. The radar has a lookdown/shootdown capability, the associated right-hand MFD in the cockpit displaying all relevant target data acquired by a FLIR, weapons sensors and the radar itself.

An initial order for five prototype and 30 production-standard JAS 39A single-seaters was supplemented in 1992 by an order for a further 110 including 14 JAS 39B two-seaters.

Specification

Powerplant: one 80.5 kN (18,100 lb st) General Electric/Volvo Flygmotor RM12 (F404-GE-400) afterburning turbofan

Dimensions: length 14.10 m (46 ft 3 in); height 4.50 m (14 ft 9 in); wing span 8.40 m (27 ft 6¾ in)

Weights: empty 6,622 kg (14,600 lb); MTOW (with external stores) approx 12,500 kg (27,560 lb)

Performance (estimated): max level speed ('clean' at 10,975 m (36,000 ft)) 2,126 km/h (1,321 mph); take-off/landing distance approx 800 m (2,625 ft)

Armament: one Mauser BK27 27 mm cannon; up to 6,500 kg (14,330 lb) of Rb15F and Rb75 ASMs, DWS 39 munitions dispensers, air-to-surface rockets, free-fall/retarded bombs, Rb74 and AIM-120 AAMs, recce/sensor pods, auxiliary fuel tanks

SEPECAT Jaguar

French Jaguars are now offered with the Exocet missile

In the early 1960s the British and French were working simultaneously on the first stages of projects for a single-seat attack warplane and two-seat operational trainer respectively, and then decided that considerable advantage could accrue from a collaborative programme for this supersonic type. The starting point was the Breguet Br.121 design, from which was evolved the Jaguar with a high-set wing and long landing gear legs for the ease loading of large weapons, and powerplant of two Adour turbofans for the high thrust/weight ratio that would provide STOL performance, as well as good flight performance. The type first flew in September 1968 and was then developed in two streams for French and British requirements. The French models are the

Jaguar A single-seat attack and Jaguar E two-seat
trainer variants, built to the extent of 160 and 40
aircraft with the 32.49kN (7,305lb st) Adour Mk
102.

The British equivalents, built to the extent of 165
and 38 aircraft respectively, were the Jaguar S (Jaguar
GR.Mk 1) and Jaguar B (Jaguar T.Mk 2) with a more
advanced nav/attack system that was later upgraded in
the single-seat aircraft to create the Jaguar GR.Mk 1A
that was used effectively in the 1991 US-led war
against Iraq and is to be upgraded with provision for
the TIALD (Thermal Imaging and Laser Designation)
pod. anti-ship missile.

Specification (Jaguar GR.Mk 1)

Powerplant: two 35.76 kN (8,040 lb st) Rolls-Royce/Turbomeca
Adour Mk 104 turbofans
Dimensions: length 16.83 m (55 ft 2½ in); height 4.89 m (16 ft
½in); wing span 8.69 m (28 ft 6 in)
Weights: take-off ('clean') 10,955 kg (24,150 lb); MTOW
15,700 kg (34,610 lb)
Performance: max level speed at 10,975 m (36,000 ft) Mach 1.6
or 1,700 km/h (1,056 mph); service ceiling 14,020 m (46,000 ft)
Armament: two 30 mm Aden Mk 4 cannon with 150 rounds per
gun; 4,763 kg (10,500 lb) of disposable stores, including AAMs,
ASMs, anti-radar missiles, free-fall or guided bombs, cluster
bombs, dispenser weapons, rocket launchers, drop tanks and
ECM pods, carried on five (optionally seven) external hardpoints.

Shenyang J-6

The Shenyang J-6 is based on the Russian MiG-19

The Shenyang J-6 is the Chinese licence-built version of the Mikoyan-Gurevich MiG-19SF 'Farmer-C' with an armament of three 30mm cannon. Deliveries began in December 1961 and, since that time, the type has been built in large numbers and also exported with the designation F-6. The J-6 is technically obsolete, but in a close-range turning engagement is still a formidable air-combat adversary as a result of its great agility and powerful cannon armament. The J-6A (export F-6A) is the Chinese equivalent of the MiG-19PF with a fixed armament of two 30mm cannon and radar to provide limited all-weather interception capability.

The J-6B is the Chinese equivalent of the MiG-

19PM 'Farmer-D' with the two 30mm cannon supplemented by two semi-active radar-homing AAMs derived from the Soviet AA-1 'Alkali' and used in association with interception radar. The J-6C 'Farmer' is a J-6 development for the day fighter role with the brake chute relocated to a bullet fairing at the base of the rudder. The J-6Xin is a development of the J-6A with Chinese radar in a sharp-tipped radome on the splitter plate rather than Soviet radar in the inlet centerbody. The Tianjin JJ-6 (export FT-6) is a trainer development equivalent to (but not identical with) the MiG-19UTI that was developed in prototype form in the USSR but then not placed in production. The JZ-6 is the Chinese version of the MiG-19R reconnaissance aircraft.

Specification (Shenyang J-6c)

Powerplant: two 31.87 kN (7,165 lb st) Liming (LM) WP-6 (Tumanskii R-9BF-811) turbojets

Dimensions: length (excluding probe) 14.90 m (48 ft 10½ in); height 3.88 m (12 ft 8½ in); wing span 9.20 m (30ft 2½ in)

Weights: take-off ('clean') 7,545 kg (16,634 lb); MTOW 10,000 kg (22,046 lb)

Performance: max level speed at 11,000 m (36,090 ft) Mach 1.45 or 1,540 km/h (957 mph); ceiling 17,900 m (58,725 ft)

Armament: three 30 mm Type 30-1 cannon; 500 kg (1,102 b) of disposable stores, including AAMs, free-fall bombs, rocket launchers and drop tanks, carried on four external hardpoints.

Shenyang J-8

It is rumoured that Iran acquired the J-8 in 1994

Close study of the Jianjiji-8 (Fighter aircraft 8) reveals traces of the MiG-21 in its design but to think of it as a MiG-21 derivative would be wrong. Developed from 1964, the first of two J-8 prototypes flew for the first time on 5 July 1969. The testing programme was then delayed by the Mao Zedong's Cultural Revolution.

The baseline production model was the J-8 which soon gave way to the J-8I, the prototype of which flew on 24 April 1981. Upgraded to operate as an all-weather fighter, the J-8I featured a Szichuan SR-4 fire control radar in the conical centrebody and adopted a 23mm two-barrel cannon in place of the J-8's 30mm weapon. Some J-8s were upgraded to J-8I standard, joining over 100 new-build aircraft in PLAAF service, but the main programme to enhance the type's capabilities with Western avionics was abandoned in 1990.

The J-8II is quite different in appearance. Gone is the nose intake and conical centrebody, replaced by a nose radome and twin lateral engine air intakes; the latter is particularly important as the more powerful WP13A II engines require a greater mass of airflow. Seven weapons stations (three per wing and one centreline) can carry a greater variety of ordnance to a total of 4,500 kg (9,921lb), and the centreline and outer underwing points are plumbed for fuel tanks.

Specification (J-8II)

Powerplant: two 65.9 kN (14,815 lb st) Liyang WP13A II afterburning turbojets

Dimensions: length 21.59 m (70 ft 10 in); height 5.41 m (17 ft 9 in); wing span 9.34 m (30 ft 7 in)

Weights: empty 9,820 kg (21,649 lb); MTOW 17,800 kg (39,242 lb)

Performance: max level speed 1,300 km/h (808 mph); max rate of climb at sea level 12,000 m (39,370 ft)/min; service ceiling 20,020 m (66,275 ft)

Armament: one Type 23-3 23 mm twin-barrel cannon with 250rds, PL-2B and PL-7 AAMs, unguided air-to-air rockets, free-fall bombs, three auxiliary fuel tanks

Sikorsky UH-60 / SH-60

A US Navy SH-60 fires a Norwegian Penguin missile

The Sikorsky S-70 was designed as successor to the Bell UH-1 helicopter and, first flying on 17 October 1974. It was ordered into production as the UH-60 Black Hawk for the US Army, and has also gained useful export orders. The army model has been built in a number of forms including the baseline UH-60A with accommodation for 14 troops or 3,629 kg (8,000lb) of freight, the MH-60A special operations model, the EH-60C electronic warfare model, the HH/MH-60G 'Pave Hawk' combat SAR/special operations helicopter, the MH-60K improved special operations model, the UH-60L upgraded basic model with T700-GE-701C engines driving through a

2,535kW (3,400shp) transmission, and the VH-60N VIP (presidential) transport model.

The type has also been adopted for maritime use. The first of these models was the SH-60B Seahawk with an avionics fit that includes search radar, MAD and ESM for use with weapons including homing torpedoes and anti-ship missiles. Next came the SH-60F Ocean Hawk designed to provide aircraft-carriers with an anti-submarine capability with dunking sonar and a torpedo armament. Produced in smaller numbers are the HH-60H combat SAR model and HH-60J Jayhawk Cost Guard SAR models.

Specification (UH-60A Black Hawk)

Powerplant: two 1,151 kW (1,560 shp) General Electric T700-GE-700 turboshafts

Dimensions: length, rotors turning 19.76 m (64 ft 10 in); height 3.76 m (12ft 4in); main rotor diameter 16.36 m (53 ft 8 in)

Weights: take-off ('clean') 7,708 kg (16,994 lb); MTOW 9,979 kg (22,000 lb)

Performance: max level speed at sea level 296 km/h (186 mph); service ceiling 5,790 m (19,000 ft)

Armament: provision for one or two 7.62 mm (0.3 in) M60 machine guns; more than 2,268 kg (5,000 lb) of disposable stores, including anti-tank missiles, mine dispensers, rocket launchers, cannon pods, machine gun pods and drop tanks, carried on the four external hardpoints of the optional External Stores Support System.

SOKO J-22 Orao/Avioane IAR-93

Romania operates nearly 200 IAR-93s in several versions

The result of collaboration between Romania and Yugoslavia, the J-22/IAR-93 is a close-support/ground attack/tactical recce aircraft with a secondary role as a low-level interceptor. Cooperation between CNIAR of Romania and Soko of Yugoslavia led to simultaneous first flights by two single-seat prototypes, one in each country, on 21 October 1974; likewise for the two two-seat prototypes on 29 January 1977. Series production started in Romania in 1979 and in Yugoslavia the following year.

The J-22 and IAR-93 carry different ordnance loads, but all have two 23mm twin-barrel cannon in the lower forward fuselage (200 rpg) and five external stores points (two under each wing and a single centreline station). The inner underwing stations are stressed for loads up to 500kg (1,102lb) and the outer two for loads up to 300kg (661lb). Some IAR-93Bs heve been known to carry up to eight AAMs, two per underwing station. Close to 200 IAR-93A/Bs have been procured by the Romanian Air Force, but production of the J-22 was halted as a result of the civil war that has torn Yugoslavia apart. Only a small number of Serb IAR-93s remain serviceable.

Specification (J-22 Orao)

Powerplant: two 22.24 kN (5,000 lb st) Turbomecanica/Orao (licence-built Rolls-Royce) Viper Mk 633-47 afterburning or 17.79 kN (4,000 lb st) Viper Mk 632-41 non-afterburning turbojets

Dimensions: length 14.90 m (48 ft 10½ in); height 4.52 m (14 ft 10 in); wing span 9.30 m (30 ft 6¼i n)

Weights: empty, equipped 5,500 kg (12,125 lb); MTOW 11,080 kg (24,427 lb)

Performance (clean, at 8,170 kg (18,012 lb)): max level speed at sea level 1,130 km/h (702 mph); max rate of climb at sea level 5,340 m (17,520 ft)/min; service ceiling 15,000 m (49,210 ft)

Armament: two 23 mm GSh-23L twin-barrel cannon with 200rpg; up to 2,800 kg (6,173 lb) of 50-500 kg (110-1,102 lb) bombs, FLAB-350 360 kg (794 lb) napalm bombs, BL755 bomblet dispensers, BRZ-127 HVAR rockets, L-57-16MD/L-128-04 rockets, 500 kg (1,102 lb) AM-500 sea mines, AGM-65B Maverick or Grom ASMs, recce pod (centreline), three 500 l (110 gal) drop tanks

SOKO G-4 Super Galeb

NATO fighters stopped Serb G-4s from bombing Bosnia

Designed as a replacement for the G-2 Galebs and
Lockheed T-33As then in service with the Federal
Yugoslav Air Force, the Super Galeb (Super Seagull)
two-seat basic/advanced trainer and ground-attack
aircraft first flew in prototype (G-4PP) form on 17
July 1978.

Three prototypes and six pre-production aircraft
were followed by the first production G-4s, these
emerging from SOKO's Mostar plant in 1983.
However, the civil war in the former Yugoslavia led to
the shelling of Mostar itself and production ceased.
Today, of the 130 or so G-4s delivered only a small
number remain operational - and they are no match
for NATO's fighters if caught in the air during

bombing missions in support of Bosnian-Serb ground forces. The sole export customer is Myanmar (Burma), 12 G-4s having been delivered before the closure of the Mostar production line.

A simple, straightforward design, the G-4 carries a respectable armament of bombs and rockets. It features the now customary twin tandem cockpit configuration, with the rear seat raised by some 25cm (10 in). Power is provided by a single licence-built Rolls-Royce Viper Mk 632-46 turbojet, fuel for which is housed in two rubber fuselage tanks, a collector tank and two integral wing tanks.

Specification (G-4)

Powerplant: one 17.8 kN (4,000 lb st) licence-built Rolls-Royce Viper Mk 632-46 non-afterburning turbojet

Dimensions: length 12.25 m (40 ft 2¼ in); height 4.30 m (14 ft 1¼ in); wing span 9.88 m (32 ft 5 in)

Weights: empty, equipped 3,172 kg (6,993 lb); MTOW 6,300 kg (13,889 lb)

Performance: (at 4,708 kg (10,379 lb)): max level speed at 4,000 m (13,120 ft) 910 km/h (565 mph); max rate of climb at sea level 1,860 m (6,100 ft)/min; ceiling 12,850 m (42,160 ft)

Armament: one ventral 23 mm twin-barrel rapid-fire cannon with 200rds; up to 1,280 kg (2,822 lb) of high-explosive bombs, S-8-18 cluster bombs, napalm pods, munitions dispensers, rocket pods, 128mm rocket pods, gun pods, two auxiliary fuel tanks

Sukhoi Su-17, Su-20 and Su-22 'Fitter'

Su-17s equipped many Warsaw Pact bomber units

Although the Su-7 'Fitter' attack fighter was highly regarded for its ruggedness it was also notable for its poor weapon load anddismal tactical radius. Sukhoi upgraded it with variable-geometry outer wing panels to improve field performance and stretch cruising range. The type entered service during 1972 in its interim Su-17 'Fitter-B' form, soon replaced by the Su-17 'Fitter-C' and definitive Su-17M 'Fitter-C' with the AL-21F-3 engine and eight hardpoints. A tactical reconnaissance model was built as the Su-17R, and the Su-20 is the export model with downgraded avionics. The Su-17M-2 'Fitter-D' has a longer, drooped nose and an upgraded nav/attack system for the delivery of nuclear weapons, and further improved variants are the Su-17M-3 'Fitter-H' and Su-17M-4 'Fitter-K', of which the latter was built up to 1990 and the end of the Su-17 programme.

Trainer models of the basic attack fighter with tandem accommodation are the Su-17UM-2 'Fitter-E' and Su-17UM-3 'Fitter-G' based respectively on the Su-17M-2. The primary export models for the operational role are the Su-22M-3 and Su-22M-4 with downgraded avionics: the former was delivered with the R-29BS-300 engine and the latter, for Warsaw Pact allies, with the AL-21F-3 engine.

Specification (Su-17M 'Fitter-C')

Powerplant: one 110,32 kN (24,802 lb st) Lyul'ka AL-21F-3 turbojet

Dimensions: length (including probe) 18.75 m (61 ft 6½ in); height 5.00 m (16 ft 5 n); wing span 13.80 m (45 ft 3 in) spread and 10.00 m (32 ft 10 in) swept

Weights: take-off ('clean') 14,000 kg (30.864 lb); MTOW 17,700 kg (39,021 lb)

Performance: max level speed at 11,000 m (36,090 ft) Mach 2.09 or 2,220 km/h (1,379 mph); service ceiling 18,000 m (59,055 ft)

Armament: two 30 mm NR-30 cannon with 70 rounds per gun; theoretical 4,500 kg (9,921 lb) but practical 1,000 kg (2,205 lb) of disposable stores, including ASMs, free-fall or guided bombs, cluster bombs, dispenser weapons, rocket launchers, drop tanks and ECM pods, carried on eight external hardpoints.

Sukhoi Su-24

Su-24 supersonic bombers have been sold to Iraq and Iran

Designed as a supersonic bomber able to deliver nuclear and conventional missiles and bombs with great precision in all weather and at low-level, the Su-24 is also used as a recce and EW platform. Impressive in size, it features a disitinctive slab-sided fuselage broad enough to accommodate two AL-21F-3A turbojets and side-by-side cockpit seating for the pilot (port side) and WSO. The shoulder-mounted wing has a maximum sweep of 60deg and features full-span leading-edge slats to enhance handling, and the large nose radome houses two radar scanners: one for nav/attack and TFR, the other for airborne ranging.

The principal bomber variant is the Su-24M. Introduced into service in 1983, the Su-24M features a shorter nose radome housing TFR (in place of the earlier terrain avoidance system) and a forward-looking attack radar. The wing roots have been extended to take two glove pylons, increasing the number of stores points to nine with a total capacity of 8,000kg (17,637lb). Laser-guided munitions can be used in conjunction with a Kaira 24 laser ranger/designator housed aft of the nosewheel door, and two AAMs can be carried for self-defence. Aircraft exported to Iraq, Iran, Syria and Libya are designated Su-24MK.

Specification (Su-24M)

Powerplant: two 109.8 kN (24,690 lb st) Saturn/Lyulka AL-21F-3A afterburning turbojets

Dimensions: length 24.59 m (80 ft 8¼ in); height 6.19 m (20 ft 3¾ in); wing span (fully swept) 10.36 m (34 ft 0 in), (fully spread) 17.64 m (57 ft 10½ in)

Weights: empty, equipped 19,000 kg (41,885 lb); MTOW 39,700 kg (87,520 lb)

Performance: max level speed at sea level ('clean') 1,320 km/h (820 mph); max rate of climb at sea level 9,000 m (29,525 ft)/min; service ceiling 17,500m (57,400 ft)

Armament: one GSh-6-23 23 mm six-barrel gun; up to 8,000 kg (17,637lb) of ordnance including TN-1000/-1200 nuclear weapons, TV/laser-guided bombs, 70 mm rockets, 23 mm gun pods, R-60 AAMs, four auxiliary fuel tanks

Sukhoi Su-25

Pilots were unhappy with the Su-25 during the Afghan war

The Su-25 was built to survive heavy punishment over the battlefield when flying in support of ground forces. It has armour protection for critical components and the pilot, the cockpit being protected by 24mm (0.9in) welded titanium. The two R-195 turbojets are housed in widely separated bays, and the internal fuel tanks are filled with reticulated foam for added protection against explosion. Low-speed handling is aided by wing-tip pods that split at the rear to form airbrakes. A flat-glass nose window covers a laser rangefinder/target designator, while at the back a Sirena-3 radar warning system antenna is located above the tailcone.

The two-seat combat-capable derivative of the

single-seat Su-25 (Su-25K for export) was developed as the Su-25UB (UBK for export), which in turn led to the unarmed Su-25UT two-seater (also known as the Su-28). A navalized version of the Su-25UT was known as the Su-25UTG. However, it was combat experience over Afghanistan that led to the most potent development, namely the Su-39. Based on the Su-25UB two-seater, the Su-39 ha a newnav/attack system with automatic weapons selection and release, wing-tip ECM pods, and a large cylindrical fairing at the base of the fin housing chaff/flare dispensers and an IR jammer.

Specification (Su-25K)

Powerplant: two 44.18 kN (9,921 lb st) Soyuz/Tumansky R-195 non-afterburning turbojets

Dimensions: length 15.53 m (50 ft 1½ in); height 4.80 m (15 ft 9 in); wing span 14.36 m (47 ft 1½ in)

Weights: empty 9,500 kg (20,950 lb); MTOW 14,600-17,600 kg (32,187-38,800 lb)

Performance: max level speed at sea level 975 km/h (606 mph) max attack speed (airbrakes open) 690 km/h (428 mph); service ceiling (with max weapons) 5,000 m (16,400 ft)

Armament: one AO-17A 30 mm twin-barrel gun with 250rds; up to 4,400 kg (9,700 lb) of air-to-ground weapons including Kh-23/-25/-29 ASMs, LGBs, S-5 57 mm rockets, S-8 80 mm rockets, S-24 240mm rockets, S-25 330 mm rockets, cluster bombs, SPPU-22 GSh-23 23 mm twin-barrel gun pods with 260rds, R-3S or R-60 AAMs, four PTB-1500 auxiliary fuel tanks

Sukhoi Su-27 'Flanker'

The Su-27M will operate from Russia's new aircraft carrier

Conceived from the late 1960s as a high-performance fighter of the relaxed-stability type with a quadruplex analog fly-by-wire control system, the Su-27 was almost certainly the USSR's first genuine look-down/shoot-down fighter with its pulse-Doppler radar and up to 10 AAMs. The first of 15 'Flanker-A' prototypes flew on 20 May 1977, and was followed in April 1981 by the 'Flanker-B' initial production model that entered service in 1984 with squared-off wing tips carrying missile launch rails. The type also has a number of aerodynamic refinements such as vertical tail surfaces located farther outboard, an extended tailcone, and leading-edge flaps. The Su-27UB 'Flanker-C' is the tandem two-seat variant first revealed in 1989 with taller vertical tail surfaces. The variant has an improved radar in a slightly longer nose, and is probably as much a combat type as a trainer, with the rear-seat officer managing of the warplane's upgraded avionics and weapon systems.

The Su-27K 'Flanker-D' is the navalized version of the 'Flanker-B' selected in 1992 for deployment on the navy's new conventional aircraft carriers. This version has folding wings and tailplane halves, a retractable flight refuelling probe, an arrester hook under a shortened bullet fairing between the paired engines, strengthened landing gear, moving canard foreplanes to allow slower landing at higher angles of attack, and engines uprated some 12% to 15%.

Specification (Su-27 'Flanker-B')

Powerplant: two 122.58 kN (27,557 lb st) Lyul'ka AL-31F turbofans

Dimensions: length (excluding probe) 21.935 m (71 ft 11½ in); height 5.932 m (1 ft 5½ in); wing span 14.70 m (48 ft 2½ in)

Weights: take-off ('clean') 22,000 kg (45,801 lb); MTOW 30,000 kg (66,138 lb)

Performance: max level speed at 11,000 m (36,090 ft) Mach 2.35 or 2,500 km/h (1,553 mph); ceiling 18,000 m (59,055 ft)

Armament: one 30 mm GSh-30-1 cannon with 149 rounds; 6,000 kg (13,228 lb) of disposable stores, including AAMs, free-fall bombs, rocket launchers, drop tanks and ECM pods, carried on 10 external hardpoints.

Sukhoi Su-35 'Flanker'

The Su-35 promises to be exceptionally manoeuverable

One of the most important new features being planned for new-generation warplanes (and also for possible retrofit on existing warplanes with a digital flight-control system able to accommodate the control law changes) is thrust-vector control. In its most complex form, this allows the thrust of the engine to be inclined in any of three planes to improve, by a very considerable degree, the aerial agility of the warplane for both offensive and defensive purposes. Current American efforts have produced vectoring in two planes, and the system is being considered for aircraft such as the forthcoming Lockheed F-22 Rapier, but it is thought that the CIS has so far limited its TVC efforts to the vertical plane.

The fighter that should benefit first from TVC is the Su-35, a development of the Su-27 scheduled to fly in 1994 for a service debut in 1996, but seriously delayed by the financial problems currently plaguing the CIS. The latest form of nozzle planned for the type allowing the thrust to be vectored about 5° up and between 10° and 15° down. Other features of the Su-35 are the Phazotron NO-11 pulse-Doppler main radar in either its Zhuk-27 slotted-array or Zhuk-PH phased-array forms, and the Phazotron NO-14 rearward-facing radar which, in concert with the MAK optronic sensor on top of the fuselage, will enhance the fighter's all-round defensive suite.

Specification (estimated):

Powerplant: two 122.58 kN (27,557 lb st) Lyul'ka AL-31F turbofans

Dimensions: length (excluding probe) 21.935 m (71 ft 11½ in); height 5.932 m (19 ft 5½ in); wing span 14.70 m (48 ft 2½ in)

Weights: take-off ('clean') 22,000 kg (45,801 lb); MTOW 30,000 kg (66,138 lb)

Performance: max level speed at 11,000 m (36,090 ft) Mach 2.35 or 2,500 km/h (1,553 mph); service ceiling 18,000 m (59,055 ft)

Armament: one 30 mm GSh-30-1 cannon with 149 rounds; 6,000 kg (13,228 lb) of disposable stores, including AAMs, free-fall bombs, rocket launchers, drop tanks and ECM pods, carried on 10 external hardpoints.

Tupolev Tu-16

Tu-16s still serve as naval strike aircraft and bombers

First flown in prototype form on 27 April 1952, the Tu-16 represented one of a new generation of Soviet bombers to emerge during the 1950s. A versatile design, some 2,000 production Tu-16s were built before production ended in 1959/60.

The first production model was the Tu-16A, a strategic bomber able to deliver up to 9,000 kg (19,842lb) of bombs. Two KS-1 turbojet-powered ASMs (one beneath each wing) were introduced on the Tu-16KS, this model also carrying a missile guidance radar in the weapons bay. The Tu-16T was a navy bomber armed with torpedoes; and anti-shipping operations were also assigned to the Tu-16K-10, the designation identifying about 100 aircraft configured to carry a K-10S ASM in an underbelly recess. These aircraft featured a much wider nose than earlier Tu-16s, a feature continued on K-10-26s armed with two underwing K-26 ASMs.

The Tu-16KS is able to carry free-fall weapons and a pair of K-11 or K-16 rocket-powered ASMs and has a new 'lump' beneath the nose housing a target acquisition radar. It was also exported to Egypt and Iraq, several Iraqi Tu-16s were destroyed during the Gulf War before they could be used in a planned chemical warfare attack.

Now in the twilight of its career, about 130 Tu-16s remain in Russian service. The H-6A bombers and H-6D maritime strike licence-built Tu-16s continue to serve with the Chinese forces.

Specification (Tu-16K-11/16)

Powerplant: two 93.05 kN (20,920 lb st) Mikulin AM-3M turbojets

Dimensions: length: 34.80 m (114 ft 2 in); height 10.36 m (34 ft 0 in); wing span 32.99 m (108 ft 3 in)

Weights: empty, equipped 37,200 kg (82,000lb); normal take-off weight 75,000 kg (165,350 lb)

Performance (at MTOW): max level speed at 6,000 m (19,700 ft) 1,050 km/h (652 mph); service ceiling 15,000 m (49,200 ft)

Armament: seven AM-23 23 mm guns in pairs (forward dorsal, rear ventral, tail turrets) and a single in the nose; up to 9,000 kg (19,800 lb) of ordnance, ASMs and various free-fall weapons

Tupolev Tu-95

Tu-95s were originally designed to bomb the USA itself

Designed to attack targets in the United States with nuclear bombs, the Tu-95 took to the air for the first time in 1952. It is still the world's only swept-wing, propeller-driven aircraft to enter service, and the four NK-12MV turboprops provide enough power to make it the fastest propeller-driven aircraft ever built. The long, thin fuselage is divided into three pressurized compartments, but the seven-man crew (two on the flight-deck, four behind in a rear-facing compartment, and a tailgunner) do not have ejection seats; a conveyor belt in the flight-deck floor carries them to an emergency exit hatch in the nosewheel door.

Production-standard Tu-95Ms entered service in 1956. Equipped to carry two nuclear bombs, they were soon joined by the Tu-95K-20 capable of carrying a Kh-20 ASM. The Kh-20 was also carried by the Tu-95KD which also introduced a nose-mounted IFR probe. The late-production Tu-95MS is based on the Tu-142's airframe, but with the Tu-95's shorter fuselage. Two versions have been built: the Tu-95MS6 carries six Kh-55 cruise missiles on an internal rotary launcher; while the Tu-95MS16 carries up to 16 of them.

The Tu-95 has also been modified for recce missions. The Tu-95RT sports an underfuselage surface search radar and a smaller undernose radar radome, rear-fuselage Elint blisters and various other antennae for recce tasks, although its primary role is that of mid-course missile guidance.

Specification (Tu-95MS)

Powerplant: four 11,033 kW (14,795 shp) KKBM Kuznetsov NK-12MV turboprops

Dimensions: length 49.50 m (162 ft 5 in); height 12.12 m (39 ft 9 in); wing span 51.10 m (167 ft 8 in)

Weights: empty, equipped 90,000 kg (198,415 lb); MTOW 188,000 kg (414,470 lb)

Performance: maximum speed at sea level 650 km/h (404 mph), at 7,620 m (25,000 ft) 925 km/h (575 mph); service ceiling 12,000 m (39,370ft), with max weapon load 9,100 m (29,850 ft)

Armament: (MS6) six Kh-55 ALCMs, (MS16) 16 Kh-55 ACLMs, single or twin 23 mm cannon in tail turret

Tupolev Tu-142

Tu-142s continue to fly across the Atlantic and Pacific

Externally very similar in appearance to the Tu-95, the
Tu-142 emerged in the late 1960s as a dedicated
maritime reconnaissance and ASW derivative for use
by Soviet Naval Aviation. The Tu-142 series is
distinguishable by a lengthened forward fuselage, a
strengthened wing incorporating double-slotted
trailing-edge flaps, more powerful Kuznetsov NK-
12MV turboprops and a strengthened 12-wheel main
landing gear. Two bays in the rear of the fuselage (one
replacing the Tu-95's rear ventral turret) are used to
house a variety of stores including sonobuoys,
torpedoes and depth charges.

Variants include the Tu-142 Mod 1 which reverted to the Tu-95's standard-size inboard engine nacelles and retained the original four-wheel main landing gear bogies. The Tu-142's prominent chin-mounted radar radome has been deleted. A further increase in the length of the forward fuselage signalled the arrival of the Tu-142M Mod 2, other features including a raised flight-deck roof and a nose-mounted IFR probe set at a slightly lower angle than on previous examples.

The sole export customer for the Tu-142 to date has been the Indian Navy, recipient of 10 Tu-142M Mod 3s which feature a new MAD blister added to the tip of the tailfin.

Specification (Tu-142M Mod 3)

Powerplant: four 11,033 kW (14,795 shp) KKBM Kuznetsov NK-12MV turboprops

Dimensions: length 49.50 m (162 ft 5 in); height 12.12 m (39 ft 9 in); wing span 51.10 m (167 ft 8 in)

Weights: empty, equipped 86,000 kg (189,594 lb); MTOW 185,000 kg (407,850 lb)

Performance: maximum level speed ('clean' at 7,620 m (25,000 ft)) 925 km/h (575 mph); service ceiling 12,000 m (39,370 ft)

Armament: twin NR-23 23 mm twin-barrel cannon in tail; up to 11,340 kg (25,000 lb) of torpedoes, nuclear/conventional depth charges, sonobuoys

Tupolev Tu-22

Tu-22s serve as bombers and electronic warfare platforms

Even before production of the Tu-16 had ended, its planned replacement had flown, in prototype form, on 7 September 1959. The Tu-22 also sported two engines, but in an unusual configuration: housed in pods located above the rear fuselage and either side of the tail fin. Other features included an all-swept mid-mounted wing featuring prominent pods extending well aft of the trailing-edge, these being used to house the main undercarriage units, and a crew of three seated in tandem, of whom only the pilot had an upward-ejecting seat. The other two ejected downwards via panels in the bottom of the fuselage!

Designed as a supersonic bomber with dash capability and a maximum load of 10,000kg

(22,046lb), the Tu-22 entered service armed with free-fall bombs housed in the underfuselage weapons bay. Overall performance of this first production model proved disappointing, and it soon gave way to the Tu-22K. Featuring a larger undernose radome housing a missile guidance radar, this model's primary weapon is the Kh-22 ASM.

An unarmed reconnaissance variant, the Tu-22R is optimized for daylight operations and features six windows in the bomb bay doors for three pairs of long-range cameras. Some Tu-22Rs have had their radar-directed 23mm tail cannon replaced by ECM equipment; others have flares and IR sensors.

Specification (Tu-22)

Powerplant: two 156.9 kN (35,275 lb st) RKBM VD-7M afterburning turbojets

Dimensions: length 42.60 m (139 ft 9 in); height 10.00 m (32 ft 9¼ in); wing span 23.50 m (77 ft 1¼ in)

Weights: basic empty 40,000 kg (88,183 lb); MTOW (with four JATO rockets) 94,000 kg (202,820 lb)

Performance: max level speed at 12,200 m (40,000 ft) 1,480 km/h (920 mph); max speed at sea level 890 km/h (553 mph); service ceiling 18,300 m (60,040 ft); service ceiling (supersonic) 13,300 m (43,635 ft)

Armament: one NR-23 23 mm gun in tail turret; up to 10,000 kg (22,046 lb) of ordnance including 24 FAB-500s and one FAB-9000 or other weapons

Tupolev Tu-22M

The Tu-22M is a 'swing-wing' development of the Tu-22

Using the same design philosophy applied to turn the Su-7 into the Su-17, the Tupolev Design Bureau turned the fixed-wing Tu-22 into the Tu-22M variable-geometry bomber capable of flying at Mach 2 at high-altitude and close to Mach 1 at low-altitude. The first series production model was the Tu-22M-2 featuring a longer-span wing, revised engine air intakes and inward-retracting main landing gear units (thus negating the need for the large trailing-edge pods). First flown in 1975, the Tu-22M-2 is powered by two NK-22 turbofans and has seats for four crewmen (pilot/co-pilot up front, navigator/WSO further aft). This model was initially armed with a single Kh-22 ASM semi-recessed beneath the fuselage, but the current configuration allows for another two

Kh-22s, one per rack located under both fixed wing centre-section panels.

Further development led to the Tu-22M-3, which features improved NK-23 turbofan engines, an upturned nose believed to house a new radar and TFR and a rotary launcher in the weapons bay able to to carry up to six Kh-15P missiles. In addition, the underwing Kh-22s can be replaced by up to four more SRAMs, and the rear-mounted gunnery has been reduced to just a single twin-barrel gun.

Some 280 Tu-22Ms have entered service, the majority going to Russian Naval Aviation units. Production continues at 20-30 aircraft per year.

Specification (Tu-22M-3)

Powerplant: two 245.2 kN (55,115 lb) Kuznetsov/KKBM NK-25 afterburning turbofans

Dimensions: length 42.46 m (139 ft 3¾ in); height 11.05 m (36ft 3in); wing span (fully swept) 23.30 m (76 ft 5½ in), (fully spread) 34.28 m (112ft 5¾ in)

Weights: basic empty 54,000 kg (119,048 lb); MTOW (with JATO rockets) 126,400 kg (278,660 lb)

Performance: max level speed at high altitude 2,000 km/h (1,242 mph); max level speed at low altitude 1,050 km/h (652 mph); service ceiling 13,300 m (43,635 ft)

Armament: one GSh-23 23 mm twin-barrel gun tail mounting; up to 24,000 kg (52,910 lb) of ordnance including Kh-22 ASMs, Kh-15P SRAMs, bombs, sea mines

Tupolev Tu-160

The Tu-160 is the world's biggest bomber

The Tu-160 was revealed to the West by a grainy satellite photograph taken a few days before the prototype made its first flight on 19 December 1981. Coincidence or not, the Tu-160 bears a remarkable resemblance to the B-1B. A slim fuselage blends into low-set variable-geometry wings via very large, smooth wingroots, and the four NK-321 turbofans are housed in podded pairs under each wing. The

four-man crew enter the aircraft via a nosewheel bay.
Fighter-type control sticks are an interesting feature in
the cockpit, but there are no high-tech displays in the
form of HUDs and MFDs to help the crew. Visual
aiming of the weapons is aided by a video camera
housed behind a fairing in the forward underfuselage.
Only 18 Tu-160s were built before production ended
in 1992, all going to the 184th Heavy Bomber
Regiment then based in Ukraine. The dissolution of
the Soviet Union has left up to 14 Tu-160s in
Ukrainian hands, although attempts are being made
to negotiate their eventual return to Russia.

Specification
Powerplant: four Samara/Trud 225 kN (50,580 lb st) NK-321
afterburning turbofans
Dimensions: length 54.10 m (177 ft 6i n); height 13.10 m (43 ft
0in); wing span (fully spread) 55.70 m (182 ft 9 in), (fully swept)
35.60 m (116 ft 9¾ in)
Weights: empty 110,000 kg (242,500 lb); MTOW 275,000 kg
(606,260 lb)
Performance: max level speed at 12,200 m (40,000 ft) 2,220
km/h (1,380 mph); max rate of climb at sea level 4,200 m
(13,780 ft)/min; service ceiling 15,000 m (49,200 ft); max
unrefuelled range 12,300 km (7,640 miles)
Armament: up to 16,330 kg (36,000 lb) of free-fall bombs, Kh-
15P SRAMs or Kh-55 ALCMs

Vought A-7 Corsair II

USAF A-7s finished 25 years' service with the Gulf War

Resulting from an urgent US Navy requirement for a subsonic medium attack warplane, the A-7 was derived aerodynamically from the F8U Crusader supersonic fighter without the latter's variable-incidence wing. The YA-7 prototype made its first flight on 27 September 1965 and soon confirmed that the type met the requirement in all but minor details. The first Corsair II production model was the A-7A, of which 197 entered service from October 1966 with a powerplant of one 50.49kN (11,350lb st) Pratt & Whitney TF30-P-6 turbofan. There followed the A-7B of which 196 were delivered with the 54.27kN (12,200lb st) TF30-P-8. Some 24 of these were later adapted as TA-7C two-seat trainers.

The A-7 was then adopted (without its name) by the US Air Force as the A-7D. The US Navy introduced these avionics and the Vulcan cannon in the A-7C, of which 67 were built. Some 49 of these were later converted to TA-7C standard, a few aircraft were later modified as EA-7L electronic warfare aircraft. The USN's definitive model was the A-7E, of which 551 were delivered, and production for the US services ended with 30 A-7K two-seat trainers for the Air National Guard, but other aircraft were built for Greece, and ex-American aircraft were later transferred to Portugal. Only these two countries now fly the type.

Specification (A-7E Corsair II)

Powerplant: one 66.72 kN (15,000 lb st) Allison TF41-A-2 turbofan

Dimensions: length 14.06 m (46 ft 1½ in); height 4.90 m (16ft ½ in); wing span 11.80 m (38 ft 9 in)

Weights: take-off ('clean') 13,154 kg (29,000 lb); MTOW 19,050 kg (42,000 lb)

Performance: max level speed at sea level 1110 km/h (690mph); service ceiling 12,800 m (42,000 ft)

Armament: one 20 mm M61A1 Vulcan six-barrel cannon with 1,000 rounds; 6,804 kg (15,000 lb) of disposable stores, including nuclear weapons, AAMs, ASMs, free-fall and guided bombs, cluster bombs, dispenser weapons, napalm tanks, rocket launchers, cannon pods, drop tanks and ECM pods, carried on eight external hardpoints.

Vought F-8P Crusader

One of the surviving Vought F-8E(FN)s

That the F-8 Crusader continues to soldier on in the service of a world leader in the field of fighter design and technology, is as much a testimony to the aircraft's longevity and fighting capabilities as it is to the French Navy's long-identified (but yet to be satisfied) need for modern fighter aircraft to operate from the decks of its two aircraft carriers, Clemenceau and Foch.

Designed to meet a US Navy requirement in the early 1950s for a carrierborne supersonic air-superiority fighter, Vought's proposal was flown in prototype form (XF8U-1) for the first time on 25 March 1955. The F-8 Crusader proved a great success in US Navy and Marine Corps service, particularly during the air war over Vietnam. Some 1,259 F-8s were built, but all have long since disappeared from US and Philippine service. Yet 30 years ago the 42nd and last F-8E(FN) was handed over to the French Navy. The F-8E(FN)s were tailored to operations from the relatively small decks of the Clemenceau and Foch. Features include two-stage leading-edge flaps, indigenous avionics matched to the R.530/Super 530 medium-range and Magic short-range AAMs, and a Magnavox APQ-94 radar. Eighteen survivors have recently been upgraded to extend their lives into the late 1990s when 20 Rafale Ms will finally take their place; their new lease of life has earned them the apt designation of F-8P (Prolonge) Crusader.

Specification

Powerplant: one Pratt & Whitney J57-P-20A afterburning turbojet

Dimensions: length 16.61m (54 ft 6 in); height 4.80 m (15 ft 9 in); wing span 10.72 m (35 ft 2 in)

Weights: empty, equipped 8,935 kg (19,700 lb); maximum loaded 15,420 kg (34,000 lb)

Performance: max speed at high altitude ('clean') 1,827 km/h (1,135 mph); initial rate of climb 6,400 m (21,000 ft)/min; typical combat radius 708 km (440 miles)

Armament: four Colt Mk 12 20 mm cannon, each with 84 or 144rds; R.530D/Super 530 medium-range AAMs, Magic/Sidewinder short-range AAMs

Westland Lynx (Army)

The Lynx is used for many roles by the British Army

Of the 13 prototypes involved in the Lynx flight testing programme, the first Army prototype took to the air on its maiden flight on 12 April 1972. This was the forerunner of the Lynx AH.1 general-purpose battlefield helicopter, 113 of which were procured by the British Army. First flown in early 1977, the Lynx AH.1's most obvious difference from its naval brethren was the use of skid-type landing gear in place of the fixed-wheel configuration. Full use was made of the cabin area, up to 12 troops being carried when the Lynx was used in the transport role. As far as armed duties were concerned, the principal weapon was the TOW ATGM, up to eight of which could be carried in two four-packs, one either side of the fuselage, plus six to eight reloads in the cabin. A roof-mounted sight was fitted above the port-side (gunner's) cockpit seat.

The Lynx AH.1 gave way to the uprated Lynx AH.7 featuring improved systems and a quieter, reverse-direction tail rotor fitted with composite blades. First flown on 7 November 1985, the first AH.7 was redelivered to the Army Air Corps on 30 March 1988. displays and lighting. The TOW-firing capability was deleted from the latest model to enter service, the Lynx AH.9, as a cost-cutting measure.

Specification (Lynx AH.7)

Powerplant: two 846 kW (1,135 shp) Rolls-Royce Gem 42-1 turboshafts

Dimensions: length 15.16 m (49 ft 9 in); height 3.50 m (11ft 6 in); width (main rotor bladed folded) 3.75 m (12 ft 3¾ in); main rotor diameter 12.80 m (42 ft 0in)

Weights: empty, equipped (anti-tank) 3,072 kg (6,772 lb); MTOW 4,876 kg (10,750 lb)

Performance (at MTOW at sea level): maximum continuous cruising speed 259 km/h (161 mph); max vertical rate of climb 472 m (1,550 ft)/min; hovering ceiling 3,230 m (10,600 m)

Armament: two 20 mm external cannon, one 7.62 mm machine gun; up to eight ATGMs, Minigun pods, rocket pods

Westland Lynx (Navy)

RN Lynxs destroyed many Iraqi missile boats in the Gulf

Although the land based Lynx AH.1 for the British Army was the first production model to enter operational service, the driving force behind the design and development of this compact helicopter was a requirement for a new naval helicopter to undertake shipborne duties.

For the Royal Navy, the first production model was the Lynx HAS.2, 60 of which were acquired to perform a variety of roles including ASW, ASV, recce, transport duties and SAR. During the 1980s all extant HAS.2s were upgraded to Gem 41-1-powered HAS.3 standard, this model introducng a Seaspray search and tracking radar in a modified nose to facilitate ASW classification and strike and ASV search and strike. Prosecution of targets comes courtesy of two homing torpedoes, four Sea Skua semi-active homing torpedoes or two depth charges. Small-scale upgrades include seven HAS.3Ss with secure speech facilities and two HAS.3ICE for deployment aboard the Antarctic survey vessel HMS Endurance. Experience in the Gulf led to 17 HAS.3GMs (Gulf Mod) with improved cooling and defensive systems. The latest HAS.8 upgrades have been incorporated in the Export Super Lynx, already acquired by Brazil, Portugal and South Korea.

Specification (Lynx HAS.3)

Powerplant: two 835 kW (1,120shp) Rolls-Royce Gem 41-2 turboshafts
Dimensions: length (both rotors turning) 15.16 m (49 ft 9 in); height 3.48 m (11ft 5in); width (main rotor blades folded) 2.94 m (9 ft 7¾ in); main rotor diameter 12.80 m (42 ft 0 in)
Weights: empty, equipped (ASW) 3,343 kg (7,370 lb); MTOW 4,763 kg (10,500 lb)
Performance (at normal MTOW at sea level): max continuous cruising speed 232 km/h (144 mph); max vertical rate of climb 351 m (1,150ft)/min; hovering ceiling 2,575 m (8,450 ft)
Armament: FN HMP 12.7 mm (0.50 in) machine-gun pod (optional); Mk44, Mk46 or Sting Ray homing torpedoes, Sea Skua semi-active homing missiles, Mk11 depth charges, marine markers

Glossary of Aviation Terms

AAA	Anti-Aircraft Artillery
AAM	Anti-Aircraft Missile
AEW	Airborne Early Warning
afterburning	burning fuel in the jet pipe to temporarily boost thrust
AGM	Air-to-Ground Missile
AGM-86	The Boeing 'Tomahawk' air launched cruise missile
AIM-7	The US 'Sparrow' radar-guided anti-aircraft missile
AIM-9	The US 'Sidewinder' infra-red homing anti-aircraft missile
ALARM	Air Launched Anti-Radiation Missile
ALCM	Air Launched Cruise Missile
AMRAAM	Advanced Medium Range Anti-Aircraft Missile
ANVIS	Aviators' Night Vision System
AN/APG	USAF designation for airborne radar systems
AoA	Angle of Attack (angle at which the airstream meets the airfoil)
Apache	French stand-off weapon dispenser
ARBS	Angle Rate Bombing Set
ARM	Anti-Radiation Missile (homes in on target's radar emissions)
ARMAT	French anti-radiation missile (Matra)
ASMP	French nuclear stand-off missile
aspect ratio	the span of a wing divided by its chord
ATAM	Air-To-Air Mistral (designation of French Gazelle helicopter armed with Mistral AAMs)
ATGM	Anti-Tank Guided Missile
avionics	aviation electronics
AWACS	Airborne Warning and Control System
BDA	Bomb Damage Assessment

BGL	French laser-guided tactical glide bomb (Matra)
bladder tank	fuel tank made from non-rigid material
BVR	Beyond Visual Range
C3	Command, control and communications
canards	foreplanes forward of the centre of gravity
canopy	transparent cockpit cover
chord	distance between the leading and trailing edges of a wing or rotor blade
CAS	Close Air Support
CAP	Combat Air Patrol
CBU	Cluster Bomb Unit (a type of bomb containing numerous small bomblets)
CCV	Control Configured Vehicle
comm/nav	communications and navigation
CKD	Component Knocked Down (i.e. for assembly elsewhere)
CFC	Carbon Fibre Composites
COIN	Counter-Insurgency (e.g. COIN Ops = Counter-Insurgency Operations)
COINS	Computer Operated Instrument System
CP/G	Co-Pilot/Gunner (second crew member in attack helicopter)
CRT	Cathode Ray Tube
dem/eval	demonstration/evaluation
DLIR	Downward-Looking Infra-Red dogtooth notch in the leading edge (also known as a sawtooth) dorsal on top of the fuselage
DVI	Direct Voice Input

ECM	Electronic Countermeasures
EFA	European Fighter Aircraft ('Eurofighter')
ELF	Extremely Low Frequency elint electronics intelligence
EMP	Electro-Magnetic Pulse (made by nuclear explosion)
ESM	Electronic Support Measures
EW	Electronic Warfare
external stores	loads, such as fuel tanks or missiles, carried outside the aircraft
flap	surface mounted on the trailing edge of a wing to increase lift during take-off/landing
FBW	Fly-By-Wire (flight surfaces controlled electronically, not mechanically)
fenestron	helicopter tail rotor with lots of thin blades rotating in a duct
FFAR	Free Flight Aircraft Rocket
FLIR	Forward-Looking Infra-Red
FSD	Full-Scale Development
ft	feet
fuselage plug	an additional section lengthening the fuselage to accommodate new equipment
G	acceleration in units of gravity
gal	gallon
gallon	1 UK gallon = 4.54 litres; 1 US gallon = 3.78 litres
GPS	Global Positioning System
GPWS	Ground Proximity Warning System
hardpoints	pylons or other fittings enabling missiles or other loads to be attached

HDD	Head Down Display
HeliTOW	Helicopter-launched TOW missile
HMD	Helmet-Mounted Display
HMS	Helmet-Mounted Sight
HOT	Euromissile anti-tank missile (Haute Subsonique Optiquement Téléguide Tiré du'un Tube)
HOTAS	Hands-On Throttle and Stick
HSOS	Helicopter Stablised Optical Sight
HUD	Head-Up Display
HUDWAC	Head-Up Display Weapons Aiming Computer
IFF	Identification Friend or Foe
IFR	Instrument Flight Rules (as opposed to Visual Flight Rules)
in	inches
INS	Inertial Navigation System
IOC	Initial Operational Capability
IR	Infra-Red
IRLS	Infra-Red Line Scan (creates TV-type image from thermal sensors)
IRST	Infra-Red Search and Track
IRCM	Infra-Red Countermeasures
JASDF	Japanese Air Self-Defence Force (air force)
J-STARS	Joint Surveillance Target Attack Radar System (in Boeing E-8)
km/h	kilometres per hour
kN	kilo Newton (1 N accelerates 1 kg of mass 1 metre/second; 1 lbf = 4.44 kN)
kW	Kilowatt

knot	1 nautical mile per hour
KIAS	Knots Indicated Air Speed
LABS	Low Altitude Bombing System
LANTIRN	Low Altitude Navigation and Targeting Infra-Red, Night
lbf	pounds of thrust
LCD	Liquid Crystal Display
LED	Light Emitting Diode
LOH	Light Observation Helicopter
loiter	maximum endurance flight
longerons	fore and aft structural members in fuselage
Loran	Long-Range Navigation
LOS	Line Of Sight
low observables	materials designed to make an aircraft harder to detect by radar, IR or any other sensor
LGB	Laser-Guided Bomb
LGW	Laser-Guided Weapon
LLTV	Low Light Television
LLLTV	Low Light Level Television
LRMTS	Laser Marked Target Seeker
m	metre(s)
M or Mach No.	ratio of the speed of sound (340 m/s; 1116 ft/sec)
MAC	US Air Force Military Airlift Command
MAD	Magnetic Anomaly Detector (detects presence of submarines underwater)
mast-mounted	Fitted to the mast, above the rotor
MFD	Multi-Function Display
Mica	French AAM produced by Matra

mini-gun	multi-barrel machine-gun capable of very high rates of fire
Mistral	French infra-red homing surface-to-air missile
monocoque	type of aircraft fuselage in which all or most of the loads are taken by the skin
MMS	Mast-Mounted Sight
mph	miles per hour
MPLH	Multi-Purpose Light Helicopter
MTOW	Maximum Take-Off Weight
NOE	Nap-Of-the-Earth (very low level flight)
Ns	Newton-second (1 N thrust applied for 1 second)
NVG	Night Vision Goggles
oleo	hydraulic leg of an aircraft's undercarriage
optronics	combined optical/electrical viewing or sighting systems
OTO Melara	Italian defence company
OTH	Over The Horizon
payload	mission-related cargo e.g. bombs, missiles, extra fuel tanks, gun or sensor pods
PGM	Precision Guided Munitions
PLAAF	Peoples' Liberation Army Air Force (Communist China)
PNVS	Passive Night Vision System
port	left
radio calibration	establishing the limitations of radio equipment e.g. reception range

radius	distance an aircraft can fly from and return to the same base
radome	dome covering a radar aerial
RAAF	Royal Australian Air Force
RAF	Royal Air Force
RATO	rounds of ammunition
rpg	rounds per gun
RWR	Radar Warning Receiver (alerts pilot to enemy radar)

s	seconds
sawtooth	a notch in the leading edge of a wing (also known as dogtooth)
SAAF	South African Air Force
SADF	South African Defence Force
SAM	Surface-to-Air Missile
SAR	Search And Rescue
SCAS	Stability and Control Augmentation System
semi-active	homing on to radiation reflected from a target, beamed from another source
service ceiling	height at which maximum rate of climb is 100 ft (30.48 m) per second
sigint	signals intelligence
signature	radar/IR/electromagnetic 'fingerprint' created by aircraft, vehicle or vessel
SLAR	Side-Looking Airborne Radar
SNEB	munitions now produced by TBA (Thomson Brandt Armament)
slat	section of the leading edge that moves forward, creating a gap between it and the wing, used to increase lift at low speeds

sonobuoy	buoy containing sonar, dropped by ships or aircraft to detect submarines
SRAM	Short-Range Attack Missile
starboard	right
Stinger	US infra-red homing surface-to-air missile
STO	Short Take-Off
STOL	Shot Take-Off and Landing
store	any object carried as an external load from pylons or hardpoints
strakes	small projections running lengthwise from the fuselage, affecting local airflow
sweepback	backwards inclination of the wing
t	tonne (1000 kg)
TACAN	Tactical Air Navigation (UHF navaid)
TADS	Target Acquisition and Detection Sight
taileron	left and right tailplanes used as control surfaces
tailplane	main horizontal tail surface
TFR	Terrain-Following Radar
ton	Imperial ton (=2,240 lb or 1,016 kg) or US ton (=2,000 lb or 907 kg)
TOW	The US BGM-71 series anti-tank missile (Tube-launched Optically-tracked Wire-guided)
transparencies	cockpit canopies
transceiver	radio receiver/transmitter
turbofan	gas turbine engine in which a large diameter fan in a short duct generates thrust
turbojet	gas turbine in which the exhaust gases deliver the thrust
turboprop	gas turbine used to drive an aircraft's propeller

UHF	Ultra High Frequency
usable fuel	mass of fuel consumable in flight, usually about 95 per cent of total capacity
useful load	usable fuel plus payload
USAF	United States Air Force
USMC	United States Marine Corps
ventral	the underside of the fuselage
vortex generators	small blades fitted to wing surfaces, modifying airflow and improving control
VFR	Visual Flight Rules (as opposed to Instrument Flight Rules)
VLF	Very Low Frequency
V/STOL	Vertical/Short Take-Off and Landing
VTOL	Vertical Take-Off and Landing
wing area	total area of wing
wing loading	aircraft weight divided by wing area
WSO	Weapons Station Officer
zero/zero seat	ejector seat designed to function even when the aircraft is stationary on the ground

COLLINS GEM
BABIES' names
a mine of information

COLLINS GEM
BEER
a mine of information

COLLINS GEM
BIRDS
a mine of information

COLLINS GEM
CALORIE Counter
a mine of information

COLLINS GEM
FACT FILE
a mine of information

COLLINS GEM
FENG SHUI
a mine of information

COLLINS GEM
FLAGS
a mine of information

COLLINS GEM
Healthy EATING
a mine of information

COLLINS GEM
QUOTATIONS
a mine of information

COLLINS GEM
SAS Self-Defence
a mine of information

COLLINS GEM
SAS Survival Guide
a mine of information

COLLINS GEM
SEASHORE
a mine of information

COLLINS GEM
TREES
a mine of information

COLLINS GEM
Understanding DREAMS
a mine of information

COLLINS GEM
WILD flowers
a mine of information

COLLINS GEM
WINE Dictionary
a mine of information